Applied Radiation Physics

The Author

Dr L. K. Mishra obtained his Post-Graduate and Doctorate degree in Biochemistry from Narendra Deva University of Agriculture and Technology, Faizabad, Uttar Pradesh. He started his academic career as Assistant Professor in Biochemistry at ND University of Agriculture and Technology Faizabad. He later served as Assistant Professor at V.B.S. Purvanchal University Jaunpur, U.P. Presently he is working as Assistant Professor (Biochemistry) at Central Agricultural University, Imphal, Manipur. He is involved in teaching Biochemistry for the last fourteen years. The author has also handled research projects funded by different agencies of the Government of India. The focus of these projects is metabolic profiling of the indigenous food products of the North East India and development of new products utilizing the biodiversity of the NE India. He has authored several research papers and review articles in International and National Journals of repute and is associated with many national and international Societies in the field of Plant Biochemistry. He worked as DBT Overseas Visiting Associate at Department of Plant Sciences, North Dakota State University, Fargo, United States. The focus of the research at North Dakota State University was exploration of phenolic linked antioxidant activities of food systems for health promotion and combating metabolic syndromes like type-2 diabetes, hypertension. In recognition of his work in the field of medicinal plants and agricultural research and development he has been awarded fellowship for the year 2019 by International Society of Noni Science, Chennai.

Applied Radiation Physics

Author
Dr L. K. Mishra

Kruger Brentt
Publishers

2024

Kruger Brentt Publishers UK. LTD.
Company Number 9728962

Regd. Office: 68 St Margarets Road, Edgware, Middlesex HA8 9UU

Library of Congress Cataloging-in-Publication Data

ISBN 978-1-78715-015-7 (Hardbound)

For information on all our publications visit our website at http://krugerbrentt.com/

Preface

Radiation physics is the emission or transmission of energy in the form of waves or particles through space or through a material medium. Radiation physics includes electro-magnetic radiation such as radio waves, visible light, and X-rays, particle radiation such as alpha, beta and neutron radiation and acoustic radiation such as ultrasound, sound, and seismic waves. Radiation may also refer to the energy, waves, or particles being radiated. Radiation comes in different forms of energy in motion. Doses of radiation and the area of interest are important considerations when imaging patients, particularly during percutaneous procedures.

Treatment of materials and products with radiation in order to modify their physical, chemical and biological properties is defined as radiation processing of materials. Radiation processing can be controlled and used for the development of the novel materials and products with desirable properties. The knowledge of the basic radiation physics, including the structure of matter, elements of nuclear physics, the nature of electromagnetic radiation, and radiation interaction with matter is required to understand irradiation processing and its potential in material sciences.

Radiation is classified in two main categories, non-ionizing and ionizing radiation depending on its ability to ionize matter. Direct ionizing radiation corresponds to the energy deposition in the material by energetic charged particle which have Coulomb interaction with an orbital electron of a target atom. Indirect ionizing radiation is realized in two steps. First, fast charged particles (electrons and positrons) are released in the material due to the photon energy deposition or due to kinetic energy deposition by neutrons, protons or heavier ions. Second, the released charged particles deposit their energy directly in the material through Coulomb interactions between these particles and orbital electrons of an atom.

Many different types of interactions can take place when radiation strikes an object. For instance, atoms in an irradiated object are neutral; they each consist of a positively charged nucleus (made up of protons and neutrons) surrounded by negatively charged electrons. The process of removing an orbital electron from an atom is called ionization. Some types of radiation can transfer energy in a manner that creates ionization in the object. X-rays and gamma rays are particles called photons that can create ionization. Microwaves, ultraviolet radiation, visible light, and infrared are also photons, but they do not result in ionization and are referred to as nonionizing radiation. Ionization created by radiation in living systems can have unique biologic consequences that are different from those caused by nonionizing radiation. RECA is related specifically to diseases found to have an association with exposure to ionizing radiation.

The present book "Applied Radiation Physics" includes many aspects of radiation physics including Introduction to electro-magnetic radiation, sources of radiation, basic of particle radiation, glimpse of acoustic radiation, interaction of charged particles, photons and neutrons, basic of detectors, radiotherapy and imaging and radiation protection. It also includes numerous worked examples to solve the problems.

It is hoped that this book will be greater use for the students of radiation science, medical physics, radiation safety and radiation dosimetry. It may also be of interest to professionals of various aspects of medical physics, biology and chemistry as well as on many branches of physics.

Suggestions to improve the quality of information and presentation of the book are utmost welcome. It is out great pleasure to express my sincere thanks to the publishers Kruger Brentt Publisher, London (UK) who spread no pains in their efforts to produce real quality within a short period of time.

Dr L. K. Mishra

Contents

1

Introduction to Electro-magnetic Radiation

1.1 Introduction

Galileo Galilei and Sir Isaac Newton were concerned about gravitation, which Newton identified as fundamental force in universe. For nearly two centuries, physicists continued to believe that there was only one type of force. Yet, as scientists became increasingly aware of molecules and atoms, anomalies began to arise in particular, the fact that gravitation alone could not account for the strong forces holding atoms and molecules together to form matter.

In 1785, French physicist/inventor Charles Coulomb established the basic laws of electrostatics and magnetism and maintained that there is an attractive force that, like gravitation, can be explained in terms of the inverse of the square of the distance between objects. That attraction itself, however, resulted not from gravity, but from electrical charge, according to Coulomb.

Few years later, Johann Karl Friedrich Gauss developed a mathematical theory for finding the magnetic potential of any point on Earth along with Danish physicist Hans Christian Oersted who became first scientist to establish the existence of clear relationship among electricity and magnetism which led to foundation of electromagnetism, devoted in study of electrical and magnetic phenomena.

French mathematician and physicist André Marie Ampère concluded that magnetism is the result of electricity in motion, and, in 1831, British physicist and chemist Michael Faraday published his theory of electromagnetic induction. This theory shows how an electrical current in one coil can set up a current in another through the development of a magnetic field. This enabled Faraday to develop the first generator, and for the first time in history, humans were able to convert mechanical energy systematically into electrical energy.

Number of contributors; but, no one had developed a unified theory explaining the relationship between electricity and magnetism, as James Clerk Maxwell in his publication, outlined the theory of electromagnetic force as a total force on an electrically charged particle, which is a combination of forces due to electrical and/ or magnetic fields around the particle.

Maxwell had thus discovered a type of force in addition to gravity, and this reflected a "new" type of fundamental interaction, or a basic mode by which particles interact in nature. Newton had identified the first, gravitational interaction, and in the twentieth century, two other forms of fundamental interaction like strong nuclear and weak nuclear.

Energy can travel by conduction or convection, two principal means of heat transfer. But the energy Earth receives from Sun gets conveyed through electromagnetic spectrum which can be transferred by radiation. Whereas conduction of convection can only take place where there is matter, which provides a medium for the energy transfer, radiation requires no medium. Thus, electromagnetic energy passes from the Sun to Earth through the vacuum of empty space.

Electromagnetic (EM) radiation is the movement of energy - through space or a medium - composed of both electric and magnetic waves. These waves oscillate at frequencies that can vary from a few cycles per second (hertz) to more than 10^{20} hertz. The electric and magnetic waves have oscillations that are oriented 90 degrees from each other and 90 degrees from the direction of travel of the wave. The wavelength is defined as the physical difference - measured in meters - between two successive positive (or negative) wave peaks. The wave travels through free space at the speed of light, so wavelength is related to frequency (Figure 1.1).

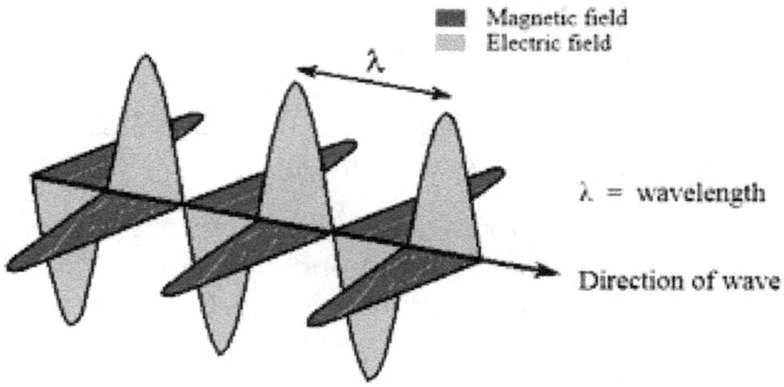

Figure 1.1: Electromagnetic Waves.

The combination of electrical and magnetic waves that operate in harmony in an EM wave contain energy. In the early days of AM radio, listeners used to buy or build 'crystal sets.' Here's a picture of a typical early set minus the headphones (which were connected to the visible posts in front) and the antenna (usually a very long wire strung outside). A crystal radio was a simple affair, consisting of an antenna, a coil, a capacitor, headphones, and a crystal of galena, germanium or other semi-conductor material.

1.1.1 Uses

EM radiation has many effects and uses in everyday life. As mentioned above, the radio band is used extensively for communications of all kinds. The Ultra-High Frequency (UHF) band, ranging from 300 megahertz (MHz) to 3,000 MHz is used mainly for communication with guided missiles, in aircraft navigation, radar, and in the transmission of television. FM radio stations use the Very High Frequency (VHF) band from 30 MHz to 300 MHz. Short wave radio uses the High Frequency (HF) band from 3 MHz to 30 MHz because waves in this band are easily reflected by the Kennelly-Heaviside layer (the E-layer) of the ionosphere, allowing very long distance communication by short wave radio. AM radio broadcasts use the Medium, Low and Very Low Frequency (MF, LF, VLF) bands from 3000 kHz down to 3 kHz. The ionosphere also reflects these waves. Note that the exact allocation of frequency bands varies from country to country and is usually controlled by government authorities.

Radio waves can be detected by a combination of:

- An aerial for receiving the electromagnetic waves and converting them into electrical oscillations
- Diodes in appropriately tuned electronic circuits in the receiver that produce an audio-frequency signal.

Microwaves, which occupy the very top of the radio wave band from 3GHz up to 300 GHz, can pass through the ionosphere and are used in radar, space communication such as with satellites, radio and television, meteorology, microwave landing system (MLS) for aircraft, distance measuring, materials research and even ordinary old cooking. Microwaves can be detected using a waveguide. This is a hollow conducting tube containing a dielectric (insulator) and is used to guide UHF EM waves along its length by reflection off the internal walls. A cavity resonator may be added to collect the energy.

Infrared radiation is heat radiation and is used in guidance systems of missiles, for linking computers in networks, as a diagnostic tool in medicine (thermography), in remote sensing aerial and satellite IR photography to search for minerals or monitor crops, in night-vision goggles, in cooking, heating, drying and so on. IR can be detected by a thermopile or a photo transistor.

Visible light is the means by which we view the world, mainly by reflection. It is also used in communication to transport huge volumes of information over very large distances by internal reflection of light in optical fibres. Light waves have high frequencies and the information-carrying capacity of a signal increases with frequency, making light perfect for the job. Light is detected by our eyes, by photo cells, cameras and light sensitive diodes.

Ultraviolet radiation is largely responsible for damage to skin and eyes exposed to sunlight for too long. It is used in the treatment of skin complaints, for killing bacteria, for fluorescent lighting, in burglar alarms, automatic door openers and counters and a host of other applications. UV radiation can be detected by photographic film, photo-voltaic cells and by the fluorescence it causes in ZnS and other salts.

X-rays are used in medicine both to supply images of internal body structures and to destroy tumors, in industry for detecting cracks in metal and in research laboratories for determining crystal structure by diffraction. X-rays can be detected by photographic plates and film, ionization of gases and by the photoelectric effect, where the X-rays knock electrons out of a metal surface.

1.2 Electromagnetic Waves and Wave Particles

The interaction among electromagnetic radiation and electromagnetic force is far obvious as even today, few people understand that there is a clear relationship between electricity and magnetism. The breakthrough in establishing that connection can be attributed both to Maxwell and to German physicist Heinrich Rudolf Hertz.

Maxwell had suggested that electromagnetic force carried with it, a certain wave phenomenon, and predicted that these waves traveled at a certain speed. In his Treatise on Electricity and Magnetism, he predicted that the speed of these waves was the same as that of light with 186,000 mi (299,339 km) per second and theorized that electromagnetic interaction covers not only electricity and magnetism, but also light. Few years later, while studying the behavior of electrical currents, Hertz confirmed that Maxwell's proposition in terms of wave phenomenon shows that an electrical current generates some sort of electromagnetic radiation.

Hertz found that the flow of electrical charges will be affected by light under certain conditions. Ultraviolet light had already been identified, and Hertz shone an ultraviolet beam on the negatively charged side of a gap in a current loop. This made it easier for an electrical spark to jump the gap. Hertz could not explain this phenomenon, which came to be known as the photoelectric effect. Indeed, no one else could explain it until quantum theory was developed in the early twentieth century. In the meantime, however, Hertz's discovery of electromagnetic waves radiating from a current loop led to the invention of radio by Italian physicist and engineer Guglielmo Marconi and others.

1.2.1 Light is Waves or Particles

At a certain point, it is required to know from the history about progression of scientists understanding about light. With advancement in area, it took over long period of time around first millennium A.D. that Alhasen described that light comes from the Sun and other self-illuminated bodies, as it had been believed up to that time from the eye itself. Hence, studies in optics, or study of light and vision, were compared to the understanding of electromagnetism, itself relatively advanced by 1666, when Newton found the spectrum of colors in light. As Newton showed, colors are arranged in a sequence, and white light is a combination of all colors.

Newton put forth the corpuscular theory of light which is an idea that light is made up of particles whereas his counterpart Christian Huygens maintained that light appears in form of a wave. For next century, adherents of Newton's corpuscular theory and of Huygens's wave theory continued to disagree. Physicists on the European continent began increasingly to accept wave theory, but corpuscular theory remained strong in Newton's homeland.

Thus, it was ironic that the physicist whose work struck the most forceful blow against corpuscular theory was himself an Englishman: Thomas Young, who in 1801 demonstrated interference in light. Young directed a beam of light through two closely spaced pinholes onto a screen, reasoning that if light truly were made of particles, the beams would project two distinct points onto the screen.

Light is a Particle (Sir Isaac Newton)

Known for his Law of Universal Gravitation, physicist Sir Isaac Newton (1643 to 1727) realized that light had frequency-like properties when he used a prism to split sunlight into its component colors. Nevertheless, he thought that light was a particle because the periphery of the shadows it created was extremely sharp and clear.

Light is a Wave (Grimaldi and Huygens)

The wave theory, which maintains that light is a wave, was proposed around the same time as Newton's theory. In 1665, Italian physicist Francesco Maria Grimaldi (1618 to 1663) discovered the phenomenon of light diffraction and pointed out that it resembles the behavior of waves. Then, in 1678, Dutch physicist Christian Huygens (1629 to 1695) established the wave theory of light and announced the Huygens 'Principle.

Light is a Wave: An Electromagnetic Wave (Maxwell)

The next theory was provided by the brilliant Scottish physicist James Clerk Maxwell. In 1864, he predicted the existence of electromagnetic waves, the existence of which had not been confirmed before that time, and out of his prediction came the concept of light being a wave, or more specifically, a type of electromagnetic wave. Until that time, the magnetic field produced by magnets and electric currents; and the electric field generated between two parallel metal plates connected to a charged capacitor were considered to be unrelated to one another. Maxwell changed this thinking when, in 1861, he presented Maxwell's equations: four equations for electromagnetic theory that shows magnetic fields and electric fields are inextricably linked. This led to the introduction of the concept of electromagnetic waves other than visible light into light research, which had previously focused only on visible light (Figure 1.2).

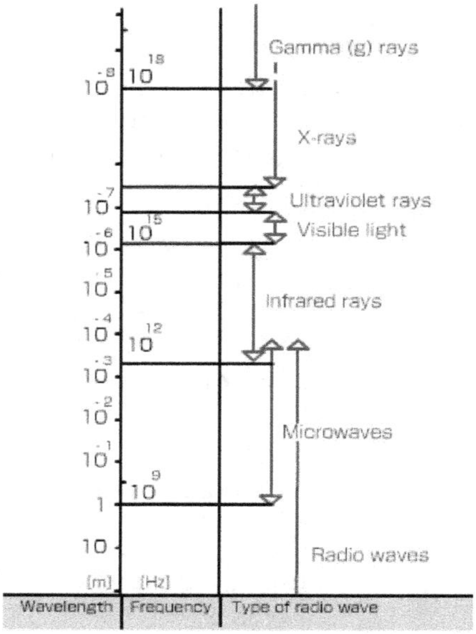

Figure 1.2: Electromagnetic Waves.

The term electromagnetic wave tends to bring to mind the waves emitted from cellular telephones, but electromagnetic waves are actually waves produced by electricity and magnetism. Electromagnetic waves always occur wherever electricity is flowing or radio waves are flying about. Maxwell's equations, which clearly revealed the existence of such electromagnetic waves, were announced in 1861, becoming the most fundamental law of electromagnetic. These equations are not easy to understand, but let's take an in-depth look because they concern the true nature of light.

Greek scientists from the ancient Pythagorean discipline postulated that every visible object emits a steady stream of particles, while Aristotle concluded that light travels in a manner similar to waves in the ocean (Figure 1.3). Even though such ideas have undergone many changes and a significant degree of evolution over past 20 centuries, the essence of which was established by Greek philosophers.

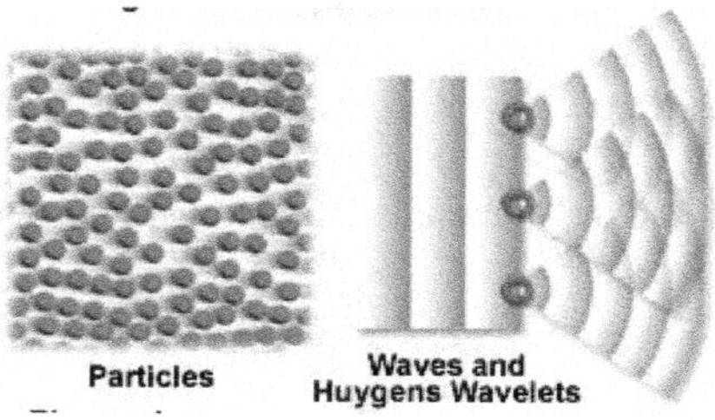

Figure 1.3: Light as Particle and Wave

One point of view envisions light as wave-like in nature, producing energy that traverses through space in a manner similar to the ripples spreading across the surface of a still pond after being disturbed by a dropped rock. The opposing view holds that light is composed of a steady stream of particles, much like tiny droplets of water sprayed from a garden hose nozzle. During the past few centuries, the consensus of opinion has wavered with one view prevailing for a period of time, only to be overturned by evidence for the other. Only during the first decades of the 20th Century was enough compelling evidence that gives comprehensive answer making both the theories to be correct.

1.2.2 Wave Motion and Electromagnetic Waves

The particle behavior of electromagnetic energy is beyond the scope as though the aspects of it, as it is necessary only to view the electromagnetic spectrum as a series of waves, and in the paragraphs that follow, the rudiments of wave motion will be shown in a short form. A type of harmonic motion that carries energy from one place to another without actually moving any matter, wave motion is related to oscillation, harmonic and typically periodic motion in one or more dimensions. Oscillation involves no net movement, but only movement in place; yet individual waves themselves are oscillating, even as the overall wave pattern moves (Figure 1.4).

It is observed that periodic motion or movement repeated at regular intervals are known as periods, that shows the behavior of periodic waves where a uniform series of crests and troughs follow each other in regular succession. Periodic waves are divided into longitudinal and transverse waves, the latter being waves in which the vibration or motion is perpendicular to the direction in which the wave is moving (Figure 1.4). Unlike longitudinal waves, such as those that carry sound energy, transverse waves are fairly easy to visualize, and assume the shape that most people imagine when they think of waves: a regular up-and-down pattern, called "sinusoidal" in mathematical terms.

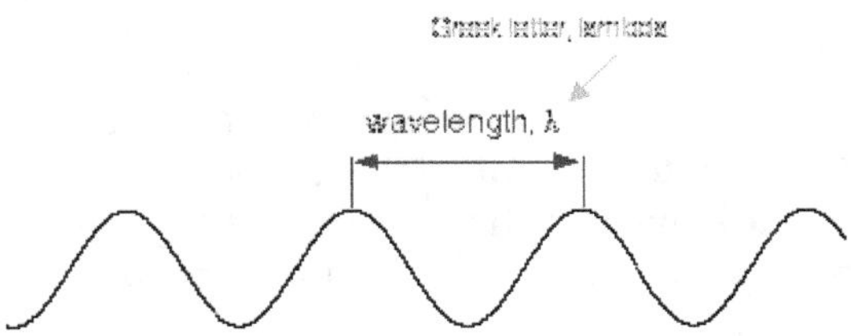

Figure 1.4: Periodic Wave Motion.

Parameters of Wave Motion

A period (T) is the amount of time required to complete one full cycle of the wave, from trough to crest and back to trough. Period is mathematically related to several other aspects of wave motion, including wave speed, frequency, and wavelength.

Frequency (f) is the number of waves passing through a given point during the interval of one second. It is measured in Hertz (Hz), named after Hertz himself: a single Hertz is equal to one cycle of oscillation per second. Higher frequencies are expressed in terms of kilohertz (kHz; 10^3 or 1,000 cycles per second); megahertz (MHz; 10^6 or 1 million cycles per second); and gigahertz (GHz; 10^9 or 1 billion cycles per second.)

Wavelength (λ) is the distance between a crest and the adjacent crest, or a trough and an adjacent trough, of a wave. The higher the frequency, the shorter the wavelength; and, thus, it is possible to describe waves in terms of either. According to quantum theory, however, electromagnetic waves can also be described in terms of photon energy level, or the amount of energy in each photon. Thus, the electromagnetic spectrum, as we shall see, varies from relatively long-wave length, low-frequency, low energy radio waves on the one end to extremely short-wave-length, high-frequency, high-energy gamma-rays on the other.

The other significant parameter for describing a wave that mathematically is independent from those so far discussed as amplitude. It is defined as maximum displacement of a vibrating material, where the amplitude is the size of a wave. Hence, greater the amplitude, the greater will be the energy the wave contains where amplitude shows intensity. The amplitude of a light wave, for instance, determines the intensity of the light.

1.2.3 Origin of Waves

Any wave is essentially just a way of shifting energy from one place to another whether the fairly obvious transfer of energy in waves on the sea or in the much more difficult-to-imagine waves in light. In waves on water, energy is transferred by the movement of water molecules. But a particular water molecule doesn't travel all the way across the Atlantic - or even all the way across a pond. Depending on the depth of the water, water molecules follow a roughly circular path. As they move up to the top of the circle, the wave builds to a crest; as they move down again, you get a trough.

The energy is transferred by relatively small local movements in the environment. With water waves it is easy to draw diagrams to show such happening using real molecules. With light it is more difficult, as the energy in light travels because of local fluctuating changes in electrical and magnetic fields, thereby producing electromagnetic radiation.

1.2.4 Wavelength, Frequency and the Speed of Light

If a beam of light in form of wave as shown is drawn without worrying too much about what exactly is causing, then the distance between two crests is called as wavelength of light (Figure 1.6).

If the numbers of crests that are passed, are counted at a particular point per second, then it shows the ***frequency*** of the light which is measured in cycles per second or in ***Hertz***, Hz. Cycles per second and Hertz mean *exactly* the same thing. It is noted that orange light, has a frequency of about 5×10^{14} Hz which means that 5×10^{14} wave peaks pass at a given point every second.

Light has a constant speed through a given substance. For example, it always travels at a speed of approximately 3×10^{8}metres per second in a vacuum. This is actually the speed that all electromagnetic radiation travels - not just visible light. There is a simple relationship between the wavelength and frequency of a particular colour of light and the speed of light:

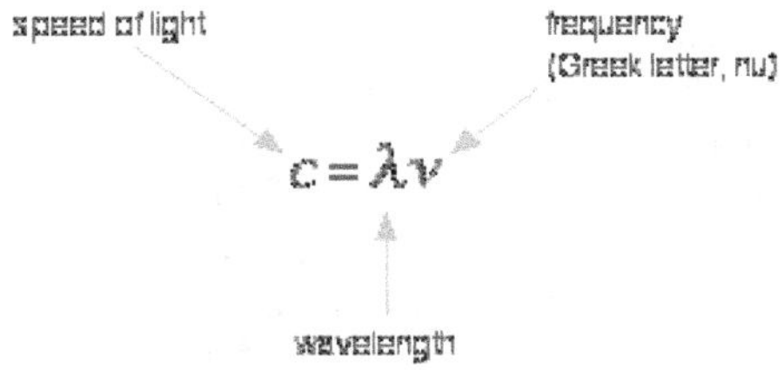

Figure 1.5: Formula of Light.

On rearranging the formula of the wavelength from given frequency and vice versa as:

$$\lambda = \frac{c}{\nu} \qquad \nu = \frac{c}{\lambda}$$

Such relationships shows that on increasing the frequency, the wavelength should decrease.

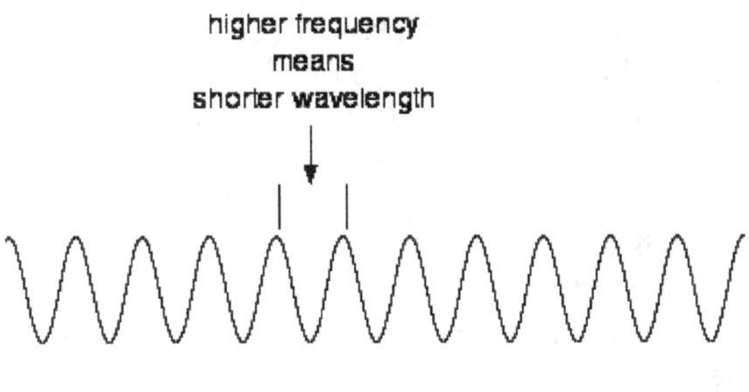

Figure 1.6: Wavelength.

Observing Figure 1.6, it is noted that the opposite is correct as; if the wavelength is long, then the frequency will be low. If two figures for the wavelengths of two different colours of light are given, then it should be observed that which one has high frequency. When, a particular colour of red light had a wavelength of 650 nm, and green light had a wavelength of 540 nm, it becomes important to know which one has higher a frequency.

As per Huygens, light waves that travel through space are mediated by ether mystical weightless substance that exists as invisible entity throughout air and space. The search for ether consumed a significant amount of resources during 19th century before finally being laid to rest. The ether theory lasted at least until the late 1800s, as evidenced by Charles Wheatstone's proposed model demonstrating that ether carried light waves by vibrating at an angle perpendicular to the direction of light propagation, and James Clerk Maxwell's detailed models describing the construction of the invisible substance. Huygens believed that ether vibrated in the same direction as light, and formed a wave itself as it carried the light waves. In a later volume, Huygens' Principle, he ingeniously described how each point on a wave could produce its own wavelets, which then add together to form a wave-front. Huygens employed this idea to produce a detailed theory for the refraction phenomenon, and also to explain why light rays do not crash into each other when they cross paths (Figure 1.7).

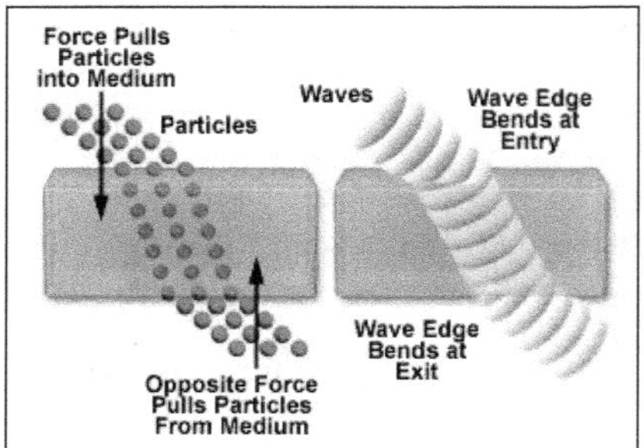

Figure 1.7: Refraction of Particles and Waves.

It is noted that when a beam of light travels between two media having different refractive indices, then the beam will undergo refraction and changes direction when it passes from the first medium into second. To find whether the light beam is composed of waves or particles, a model for each can be set to explain the phenomenon as shown in figure above. As per Huygens' wave theory, a small portion of each angled wavefront should impact the second medium before the rest of the front reaches the interface. This portion will start to move through the second medium while the rest of the wave is still traveling in the first medium, but will move more slowly due to the higher refractive index of the second medium. Because the wavefront is now traveling at two different speeds, it will bend into the second medium, thus changing the angle of propagation. In contrast, particle theory has a rather difficult time explaining why particles of light should change direction when they pass from one medium into another. Proponents of the theory suggest that a special force, directed perpendicular to the interface, acts to change the speed of the particles as they enter the second medium. The exact nature of this force was left to speculation, and no evidence has ever been collected to prove the theory.

1.3 Electromagnetic Spectrum and the Esoteric

1.3.1 Electromagnetic Spectrum

Electromagnetic wave has a transverse nature where on moving forward, it oscillates in a direction perpendicular to the line of propagation. These waves can be defined as a transverse wave with mutually perpendicular electrical and magnetic fields that emanate from it.

The electromagnetic spectrum is a complete range of electromagnetic waves on a continuous distribution from a very low range of frequencies and energy levels, with a correspondingly long wavelength, to a very high range of frequencies and energy levels, with a correspondingly shorter wavelength. Included on the electromagnetic spectrum are radio waves and microwaves; infrared, visible, and ultraviolet light; X-rays, and gamma-rays. Though each occupies a definite place on the spectrum, the divisions between them are not firm: as befits the nature of a spectrum, one simply "blurs" into another.

The electromagnetic spectrum describes a wide range of different electromagnetic waves. Also called EM waves, these are a special type of waves that can travel without a medium (Figure 1.8). Unlike sound waves and water waves, electromagnetic waves don't need a fluid, or a solid, or even air to help them travel from one place to another. EM waves can travel across the great vacuum of space, which is why we see light from distant stars and planets.

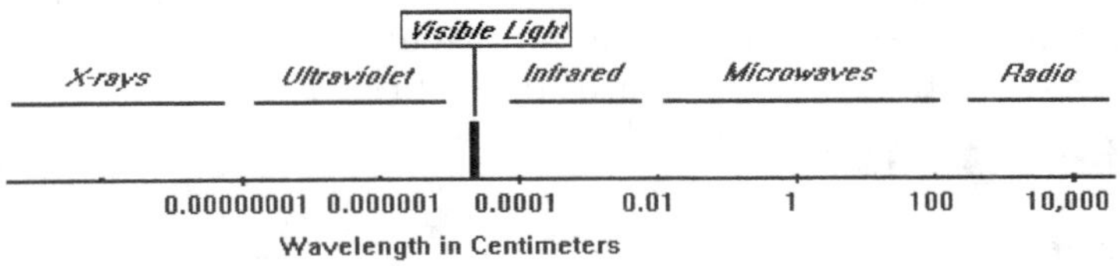

Figure 1.8: Electromagnetic Spectrum.

Electromagnetic waves are named for the fact that they have both an electric and a magnetic component. They begin when charged particles, like electrons, vibrate due to the various forces acting on them. The vibration of charged particles results in an emission of energy known as electromagnetic radiation. EM waves propagate outward from the source. Just like regular transverse waves, the oscillations of EM waves are perpendicular to the direction of the wave's travel. But, EM waves are more complicated; the electric component oscillates in one plane, while the magnetic component oscillates in a different plane. In a vacuum, EM waves always travel at the same speed - the speed of light, which is roughly 300 million meters per second. We call this value the speed of light, but really, it counts as the normal speed for all of the EM waves.

Electromagnetic waves include infrared, ultraviolet, radio waves, and microwaves. They also include X-rays and gamma-rays. Electromagnetic radiation is made when an atom absorbs energy. The absorbed energy causes one or more electrons to change their locale within the atom. When the electron returns to its original position, an electromagnetic wave is produced. Depending on the kind of atom and the amount of energy, this electromagnetic radiation can take the form of heat, light, ultraviolet, or other electromagnetic waves.

There are several many ways of causing atoms to absorb energy where excitation of atoms is done with electrical energy using neon signs. The electricity kept in neon tubes will excite or add energy to neon atoms where electrons in these atoms are in high energy state. The electrons don't like to be in the high energy state and will fall back down into the low energy state giving off radiation which we see as light.

It shows that why light is seen while neon signs appears as reddish-purple in colour. It is noted that neon signs are certain color which results when a gas is excited by electricity that tends to emit certain colors. On taking neon sign and separating the colors out with a prism we see the following spectrum (Figure 1.9):

Figure 1.9: Neon Light Spectrum.

Here it is observed that how light can produce colors. It is noted that white light comes from all the colors. On taking all the colors and combining them, white light will appear. If we place sunlight or light from an incandescent light bulb though a prism we would see the following spectrum (Figure 1.10):

Figure 1.10: Spectrum.

The above spectrum is different from the neon light spectrum as it is continuous where an entire band of light and not just several different lines. The reason why spectrum looks different is that it was not generated by electricity exciting particular gases, but was generated by heat exciting atoms. In both a light bulb and on Sun, atoms are heated to a very high temperature where the temperature is hot enough to excite atoms in giving off the light which is in form of continuous color spectrum.

Frequency Range

The range of frequencies for waves in the electromagnetic spectrum is from approximately 102 Hz to more than 10^{25} Hz. These numbers are an example of scientific notation, which makes it possible to write large numbers without having to include a string of zeros. Without scientific notation, the large numbers used for discussing properties of the electromagnetic spectrum can become bewildering.

The first number given, for extremely low-frequency radio waves, is simple enough 100, but the second would be written as 1 followed by 25 zeros. In any case, 10^{25} is a much simpler figure than 10,000,000,000,000,000,000,000,000 or 10 trillion. As noted earlier, gigahertz or units of 1 billion Hertz, are often used in describing extremely high frequencies, in which case the number is written as 1016 GHz. For simplicity's sake, however, in the present context, the simple unit of Hertz is used wherever it is convenient to do so.

Wavelength

The range of wavelengths found in the electromagnetic spectrum is from about 10^8 centimeters to less than 10^{-15} centimeters. The first number, equal to 1 million meters (about 621 mi), obviously expresses a great length. This figure is for radio waves of extremely low frequency; ordinary radio waves of the kind used for actual radio broadcasts are closer to 10^5 centimeters (about 328 ft).

For such large wavelengths, the use of centimeters might seem a bit cumbersome; but, as with the use of Hertz for frequencies, centimeters provide a simple unit that can be used to measure all wavelengths. Some charts of the electromagnetic spectrum nonetheless give figures in meters, but for parts of the spectrum beyond microwaves, this, too, can become challenging. The ultra-short wavelengths of gamma-rays, after all, are equal to one-trillionth of a centimeter. By comparison, the angstrom which is a unit that is so small to be used to measure the diameter of an atom of around 10 million times as large.

Energy Levels

Finally, in terms of photon energy, the unit of measurement is the electron volt (eV), which is used for quantifying the energy in atomic particles. The range of photon energy in the electromagnetic spectrum is from about 10^{-13} to more than 10^{10} electron volts. Expressed in terms of joules, an electron volt is equal to 1.6×10^{-19} J.

To equate these figures to ordinary language would require a lengthy digression; suffice it to say that even the highest ranges of the electromagnetic spectrum possess a small amount of energy in terms of joules. Remember, however, that the energy level identified is for a photon—a light particle. Again, without going into a great deal of detail, one can just imagine how many of these particles, which are much smaller than atoms, would fit into even the smallest of spaces. Given the fact that electromagnetic waves are traveling at a speed equal to that of light, the amount of photon energy transmitted in a single second is impressive, even for the lower ranges of the spectrum. Where gamma-rays are concerned, the energy levels are positively staggering.

Light: Invisible, Visible, and Invisible

Between 10^{13} and 10^{17} Hz on electromagnetic spectrum, there is a range of light which is infrared, visible and ultraviolet. Light actually constitutes a small portion of spectrum, and the area of visible light is very small, extending from about $4.3 \cdot 10^{14}$ to $7.5 \cdot 10^{14}$ Hz. The latter, incidentally, not only is easier not to use a string of zeroes, but where a coefficient or factor is other than multiple of 10, which is preferable to use called as significant figures which is normally a single digit followed by decimal point and up to 3 decimal places.

Infrared light lies just below visible light in frequency, and this is easy to remember because of the name: red is the lowest in frequency of all the colors. Similarly, ultraviolet lies beyond the highest-frequency color, violet.

Infrared Light

As infrared light cannot be seen, William Herschel, detected infrared radiation from Sun, that shows its existence in 1800 with thermometer. Using a prism, a 3D glass shape is used for diffusing beams of light where a beam of sunlight was directed toward the thermometer that registered heat of infrared rays.

Eighty years later, Sir William Abney developed infrared photography that captures infrared radiation instead of visible light on a film. By mid-20th century, infrared photography had come into use for many purposes that is military forces uses infrared to detect presence of enemy troops, medicine makes use of infrared photography for detecting tumors and astronomers use infrared to detect stars.

The uses of infrared imaging in astronomy, as a matter of fact, are many. The development in the 1980s of infrared arrays, two-dimensional grids which produce reliable images of infrared phenomena, revolutionized infrared astronomy. Because infrared penetrates dust much more easily than does visible light, infrared astronomy makes it easier to see regions of the universe where stars—formed from collapsing clouds of gas and dust—are in the process of developing. Because hydrogen molecules emit infrared radiation, infrared astronomy helps provide clues regarding the distribution of this highly significant chemical element throughout the universe.

Ultraviolet Light

Very little of the Sun's ultraviolet light penetrates Earth's atmosphere—a fortunate thing, since ultraviolet (UV) radiation can be very harmful to human skin. A suntan, as a matter of fact, is actually the skin's defense against these harmful UV rays. Due to the fact that Earth is largely opaque, or resistant, to ultraviolet light, the most significant technological applications of UV radiation are found in outer space.

In cooperation with several space agencies, launched International Ultraviolet Explorer, which measured UV radiation from tens of thousands of stars, nebulae, and galaxies. Despite of progress with IUE, awareness of its limitations that includes mirror of 17 inch on telescope itself was part of development.

This was the Extreme Ultraviolet Explorer (EUVE), which could observe UV phenomena over a much higher range of wavelengths than those observed by IUE. In addition, the Hubble Space Telescope, launched by the United States in 1990, includes a UV instrument called the Goddard High Resolution Spectrograph. With a mirror measuring 8.5 ft (2.6 m), it is capable of observing objects much fainter than those detected earlier by IUE.

Ultraviolet astronomy is used to study the winds created by hot stars, as well as stars still in the process of forming, and even stars that are dying. It is also useful for analyzing the densely packed, highly active sectors near the centers of galaxies, where both energy and temperatures are extremely high.

X-Rays

Though they are much higher in frequency than the visible light with wavelengths about 1,000 times shorter than for ordinary light rays are X-rays that are familiar part of modern life due to their uses in medicine. German scientist Wilhelm Röntgen developed first X-ray device and, thus, the science of using X-ray machines is called roentgenology.

The new invention became curiosity with carnivals which offers patrons, an opportunity to look at insides of their hands. With just as many people today fearing the opportunities for invasion of privacy offered by technology, many at the time worried that X-rays that allow robbers and peeping toms to look into people's houses. Soon, however, it became clear that the most important application of X-rays lay in medicine.

Gamma-rays

At the furthest known reaches of electromagnetic spectrum are gamma-rays which have ultra-high-frequency, high energy and have short wavelengths. In 1979, a wave of enormous energy passed over the Solar System, whose effects on Earth were negligible, instruments aboard several satellites, provides data concerning an enormous quantity of radiation caused by gamma-rays. As to the source of the rays themselves, believed to be a product of nuclear fusion on some other body in the universe, scientists knew nothing.

Compton Gamma-ray Observatory Satellite, launched by NASA detected number of gamma-ray bursts over the following two years. The energy in these bursts was staggering: just one of these, scientists calculated, contained more than a thousand times as much energy as the Sun will generate in its entire lifetime of 10 billion years.

Some astronomers speculate that the source of these gamma-ray bursts may ultimately be a distant supernova, or exploding star. If this is the case, scientists may have found the supernova; but do not expect to see it in the night sky. It is not known just how long ago it exploded, but its light appeared on Earth some 340,000 years ago, and during that time it was visible in daylight for more than two years.

1.3.2 Electromagnetic Waves and Esoteric

In many esoteric text, there is talk about a "vibration rate" and many believe that "higher" energy or a greater vibration rate is somehow more spiritual or more powerful than a lower vibration. Similarly, the "lower" energy or lower vibration rate is seen as not good if not evil. There also, in the chakra system some are of the belief the energy associated with the seventh chakra of the crown of the head is "higher" and more spiritual than the energy associated with the first chakra or base chakra of the groin area.

However, the study of energy as it exists in Creation does not support such conclusions and such judgment are simply that, judgments and opinions of the mind and nothing more. What needs to be understood is that traditionally any experience that was not rational or seemingly rational to the mind, was referred to some type of religious teacher, healer or spiritual teacher for they were seen as somehow knowledgeable of the unseen worlds and that which could not be explained by the rational mind. Although modern society has moved away from sole reliance on religious teachers as authorities of the unseen realms, many still consider the unseen aspects of reality, especially one's internal world as something more akin to religious or spiritual world. Additionally, until physicists started to play with unseen energy of the external world in the form of the electromagnetic spectrum and the nuclear world and psychologist and psychiatrist started to deal with the unseen inner world, all our symbols, language and the like for the unseen inner and outer realms of existence were religious. Hence, any time someone has an experience of the unseen realms whether it was inner or outer, it was seen as some type and kind of religious experience.

Now with a more complete, yet incomplete understanding, of the unseen inner and outer worlds it is clear there is greater energy in a higher vibration than a lower vibration. However, the fact on this is that, a higher frequency of vibration or greater energy says nothing about one being better or worse than a lower frequency. It all depends on what one desires to create and experience. High frequencies can be just as damaging if not more damaging to certain systems than lower vibration.

To say a higher vibration is better than a lower vibration is like saying the energy in the transformer outside our house is not the same as the energy in the wires of our house. The issue is, they have different characteristics and can do different things. The question is what is it you are trying to create. The key is to what one desires to create and then access that appropriate energy.

It needs to be realized that certain energies within a given environment are not conducive or useful for creating certain types and kinds of effect. Whereas other energies are extremely useful. This issue is not whether one energy is better than another. It is what energy is best for what one desires to create. In the oneness of Creation there is no spirit or physical. There is just what is, as it is. Nothing is any better or any worse than anything else. The question is, "Given the environment in which we find ourselves, including our body, and what we think and believe at this point in time, what experience serve us and what experiences don't?'

What needs to be understood is that electromagnetic spectrum is an infinite spectrum of energy and covers the complete range of energy. In the electromagnetic spectrum, each vibration is exactly the same except for one property. That one property is the frequency of vibration. It is different for each point on the spectrum. That difference in turn makes the vibration have a different shape within the same set of bounds. Yet each point or each frequency follows the same rules.

In many ways the electromagnetic spectrum is a very good analogy for what is described in the Creation Story for the Creativity Perspective. In the electromagnetic spectrum, each frequency of the spectrum operates just like any other point. Between any two points on the spectrum, there is an infinite number of points between an infinite number of points on either side. Those points that are near each other are seen to be the same or similar. More the two points move away from each other, more they are seen as different. Two points that are greatly removed from each other will look totally different, yet each is operating the same way following the same rules. In a similar way, each point in creation is an aspect of the whole and operates the same way as every other point and each follows the same rule. But because each point is a unique aspect of the whole, it will be seen as reacting differently in any give creation/Creation.

1.4 Real-Life Applications

1.4.1 Radio Sub-Spectrum

Among the most familiar parts of electromagnetic spectrum, in modern life is the radio. In schematic representations of spectrum, radio waves are shown either at left end or bottom with a fact that these are electromagnetic waves with low frequencies, longest wavelengths and smallest levels of photon energy. In broad sub-spectrum with frequencies of about 10^7 Hertz, are long-wave radio, short-wave radio, and microwaves. The areas of communication affected are many: broadcast radio, television, mobile phones, radar and high specific forms of technology.

As per the work of Maxwell and Hertz, foundational of radio waves for human use was practical application of radio that begins with Marconi. During 1890s, Marconi made first radio transmissions and by the end of century, he succeeded in transmitting telegraph messages across the Atlantic. Its spark transmitters sends only coded messages and with broad, long-wave length signals, only few stations could broadcast at same time. The development of electron tube in early years of 20th century made it possible to transmit narrow signals on stable frequencies which enable the development of technology for sending speech and music over airwaves.

1.4.2 Broadcast Radio

Radio signal serves as a carrier where information is added with complex sounds like speech or music named as modulation. The first modulation developed was AM or amplitude modulation, that shows first radio broadcast in 1906. Amplitude modulation varies the instantaneous amplitude of the radio wave, a function of the radio station's power, as a means of transmitting information.

By the end of World War I, radio had emerged as a popular mode of communication: for the first time in history, entire nations could hear the same sounds at the same time. During the 1930s, radio became increasingly important, both for entertainment and information. Families in the era of Great Depression would gather around large radio stations which named for their size and shape for purpose of comedy programs, soap operas, news programs and speeches.

Throughout this era, for more than half-century from end of First World War to height of the Vietnam Conflict in mid-1960s, AM held a dominant position in radio which remained the case despite of limitations that are inherent in it. AM broadcasts flickered with popping noises from lightning, for instance, and cars with AM radios tended to lose their signal when going under a bridge. Yet, another mode of radio transmission was developed in 1930s, which was FM or frequency modulation that varied the radio signal's frequency instead of its amplitude.

Not only did FM offer a different type of modulation; it was on an entirely different frequency range. Whereas AM is an example of a long-wave radio transmission, FM is on the microwave sector of the electromagnetic spectrum, along with television and radar. Due to its high frequency and form of modulation, FM offered a "clean" sound as compared with AM. The addition of FM stereo broadcasts in the 1950s offered still further improvements; yet despite the advantages of FM, audiences were slow to change, and FM did not become popular until the mid-to late 1960s.

1.4.3 Microwaves

Though microwaves are treated separately from radio waves, in fact, they are just radio signals of a very short wavelength. As noted earlier, FM signals are actually carried on microwaves, and, as with FM in particular, microwave signals in general are very clear and very strong, but do not extend over a great geographical area. Nor does microwave include only high-frequency radio and television; in fact, any type of information that can be transmitted via telephone wires or coaxial cables can also be sent via a microwave circuit.

Microwaves have a very narrow, focused beam: thus, the signal is amplified considerably when an antenna receives it. This phenomenon, known as "high antenna gain," means that microwave transmitters need not be highly powerful to produce a strong signal. To further the reach of microwave broadcasts, transmitters are often placed atop mountain peaks, hilltops, or tall buildings. In the past, a microwave-transmitting network such as NBC or CBS required a network of ground-based relay stations to move its signal across the continent. The advent of satellite broadcasting in the 1960s, however, changed much about the way signals are beamed: today, networks typically replace, or at least augment, ground-based relays with satellite relays.

1.4.4 Radio Waves for Measurement and Ranging

Radio waves can be used to send communication signals, or even to cook food; they can also be used to find and measure things. One of the most obvious applications in this regard is radar, an acronym for Radio Detection and Ranging.

Radio makes it possible for pilots to "see" through clouds, rain, fog, and all manner of natural phenomena—not least of which is darkness. It can also identify objects, both natural and man-made, thus enabling a peacetime pilot to avoid hitting another craft or the side of a mountain. On the other hand, radar may help a pilot in wartime to detect the presence of an enemy. Nor is radar used only in the skies, or for military purposes, such as guiding missiles: on the ground, it is used to detect the speeds of objects such as automobiles on an interstate highway, as well as to track storms.

In the simplest model of radar operation, the unit sends out microwaves toward the target, and the waves bounce back off the target to the unit. Though the speed of light is reduced somewhat, due to the fact that waves are traveling through air rather than through a vacuum, it is, nonetheless, possible to account for this difference. Hence, the distance to the target can be calculated using the simple formula $d = vt$, where d is distance, v is velocity, and t is time.

Typically, a radar system includes the following: a frequency generator and a unit for controlling the timing of signals; a transmitter and, as with broadcast radio, a modulator; a duplexer, which switches back and forth between transmission and reception mode; an antenna; a receiver, which detects and amplifies the signals bounced back to the antenna; signal and data processing units; and data display units. In a mono-static unit where the transmitter and receiver are in the same location, the unit has to be continually switched among sending and receiving modes.

In order to determine the range to a target whether a mountain, enemy aircraft or storm, target itself should be first detected. It is a challenge as only small portion of transmitted pulse comes back to the receiving antenna. At the same time, the antenna receives reflections from a number of other objects, and it can be difficult to determine which signal comes from the target. For an aircraft in a wartime situation, these problems are compounded by the use of enemy countermeasures such as radar "jamming." Still another difficulty facing a military flyer is the fact that the use of radar itself, the transmission of microwaves that makes the aircraft detectable to opposing forces.

Glossary

- **Amplitude:** The maximum displacement of a vibrating material. In wave motion, amplitude is the "size" of a wave, an indicator of the energy and intensity of the wave.

- **Cycle:** One complete oscillation. In wave motion, this is equivalent to the movement of a wave from trough to crest and back to trough.
- **Electromagnetic force:** The total force on an electrically charged particle, which is a combination of forces due to electrical and/or magnetic fields around the particle. Electromagnetic force reflects electromagnetic interaction, one of the four fundamental interactions in nature.
- **Electromagnetic spectrum:** The complete range of electromagnetic waves on a continuous distribution from a very low range of frequencies and energy levels, with a correspondingly long wavelength, to a very high range of frequencies and energy levels, with a correspondingly short wavelength. Included on the electromagnetic spectrum are long-wave and short-wave radio; microwaves; infrared, visible, and ultraviolet light; X-rays, and gamma-rays.
- **Electromagnetic wave:** A transverse wave with electrical and magnetic fields that emanate from it. The directions of these fields are perpendicular to one another, and both are perpendicular to the line of propagation for the wave itself.
- **Electromagnetism:** The branch of physics devoted to the study of electrical and magnetic phenomena.
- **Field:** A region of space in which it is possible to define the physical properties of each point in the region at any given moment in time.
- **Frequency:** In wave motion, frequency is the number of waves passing through a given point during the interval of one second. The higher the frequency, the shorter the wavelength. Measured in Hertz, frequency is mathematically related to wave speed, wavelength, and period.
- **Fundamental interaction:** The basic mode by which particles interact. There are four known fundamental interactions in nature: gravitational, electromagnetic, and strong nuclear, and weak nuclear.
- **Harmonic motion:** The repeated movement of a particle about a position of equilibrium, or balance.
- **Hertz:** A unit for measuring frequency, named after nineteenth-century German physicist Heinrich Rudolf Hertz (1857-1894). High frequencies are expressed in terms of kilohertz (khz; 103 or1,000 cycles per second); megahertz (mhz;106 or 1 million cycles per second); and gigahertz (ghz; 109 or 1 billion cycles per second.)
- **Intensity:** Intensity is the rate at which a wave moves energy per unit of cross-sectional area.
- **Oscillation:** A type of harmonic motion, typically periodic, in one or more dimensions.

- **Period:** For wave motion, a period is the amount of time required to complete one full cycle. Period is mathematically related to frequency, wavelength, and wave speed.
- **Periodic motion:** Motion that is repeated at regular intervals. These intervals are known as periods.
- **Periodic wave:** A wave in which a uniform series of crests and troughs follow one after the other in regular succession.
- **Photon:** A particle of electromagnetic radiation carrying a specific amount of energy, measured in electron volts (ev). For parts of the electromagnetic spectrum with a low frequency and long wavelength, photon energy is relatively low; but for parts with a high frequency and shortwave length, the value of photon energy is very high.
- **Propagation:** The act or state of traveling from one place to another.
- **Radiation:** The transfer of energy by means of electromagnetic waves, which require no physical medium (for example, water or air) for the transfer. Earth receives the Sun's energy, via the electromagnetic spectrum, by means of radiation.
- **Scientific notation:** A method used by scientists for writing extremely large numbers. This usually involves a coefficient, or factor, of a single digit followed by a decimal point and up to three decimal places, multiplied by 10 to a given exponent. Thus, instead of writing 75,120,000, the preferred scientific notation is 7.512 · 107. To visualize the value of very large multiples of 10, it is helpful to remember that the value of 10 raised to any power n is the same as 1 followed by that number of zeroes. Hence 1025, for instance, is simply 1 followed by 25 zeroes.
- **Transverse wave:** A wave in which the vibration or motion is perpendicular to the direction in which the wave is moving.
- **Wavelength:** The distance between a crest and the adjacent crest, or the trough and an adjacent trough, of a wave. Wavelength, symbolized λ (the Greek letter lambda) is mathematically related to wave speed, period, and frequency.
- **Wave motion:** A type of harmonic motion that carries energy from one place to another without actually moving any matter.

References

Atkins, Peter and Julio de Paula. Physical Chemistry for the Life Sciences. New York: Oxford University Press, 2006.

Branley, Franklyn Mansfield. The Electromagnetic Spectrum: Key to the Universe. Illustrated by Leonard D. Dank. New York: Crowell, 1979.

Chang, Raymond. Physical Chemistry for the Biosciences. USA: University Science Books, 2005.

McQuarrie, Donald and Simon, John. Physical Chemistry: A Molecular Approach. Sausalito, CA: University Science Books, 1997.

Price, Nicholas and Dwek, Raymond and Wormald, Mark. Principles and Problems in Physical Chemistry for Biochemists. R. G. Ratcliffe. New York: Oxford University Press, 1997.

Skurzynski, Gloria. Waves: The Electromagnetic Universe. Washington, D.C.: National Geographic Society, 1996.

2

Sources of Radiation

2.1 Introduction

Radiation is the energy that comes from a source and travels through space and may be able to penetrate through various materials. Light, radio, and microwaves are types of radiation that are called non-ionizing. The kind of radiation we are explaining is ionizing radiation as it can produce charged particles (ions) in matter.

Ionizing radiation is produced by unstable atoms. Unstable atoms differ from stable atoms because unstable atoms have an excess of energy or mass or both. Radiation can also be produced by high-voltage devices (*e.g.*, X-ray machines).

Atoms with unstable nuclei are said to be radioactive. In order to reach stability, these atoms give off, or emit, the excess energy or mass. These emissions are called radiation. These kinds of radiations are electromagnetic (like light) and particulate (*i.e.*, mass given off with the energy of motion). Gamma radiation and X-rays are examples of electromagnetic radiation. Gamma radiation originates in the nucleus while X-rays come from the electronic part of the atom. Beta and alpha radiation are examples of particulate radiation.

Interestingly, there is a "background" of natural radiation everywhere (ubiquitous) in our environment. Ubiquitous background radiation comes from space (*i.e.*, cosmic rays) and from naturally occurring radioactive materials contained in the earth and in living things.

After more than 80 years of intensive scientific study, radiation is the most understood, easily detected, precisely measured and strictly regulated of all environmental agents. Radiation contributed by human technology is identical to naturally occurring radiation. Most of this radiation comes from X-rays and radioactive materials used in medicine. Mining, burning fuels and buildings also contribute to radiation doses. So does fallout from past testing of nuclear weapons, color television and smoke detectors (Figure 2.1).

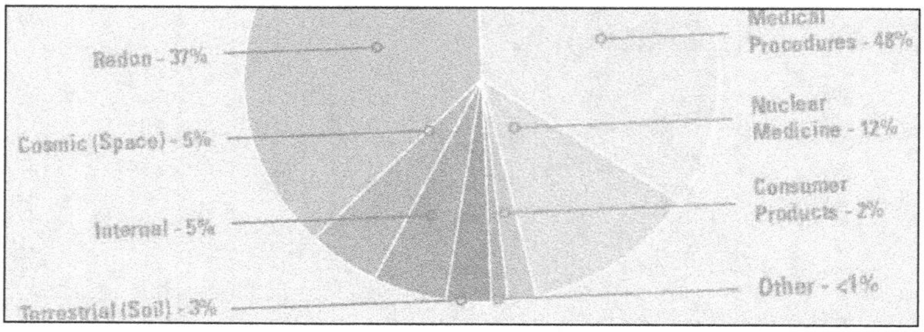

Figure 2.1: Radiation Sources.

Radiation is measured in units called millirems. The average dose per person from all sources is about 620 millirems per year. It is not, however, uncommon for any of us to receive far more than that in a given year, often due to medical procedures we may undergo. International standards allow exposure to as much as 5,000 millirems per year for those who work with and around radioactive material.

There are many sources of harmful, high energy radiation. Industrial radiographers are mainly concerned with exposure from X-ray generators and radioactive isotopes, but let's start by considering sources of radiation in general. It is important to understand that eighty percent of human exposure comes from natural sources such as outer space, rocks and soil, radon gas, and the human body. The remaining twenty percent comes from man-made radiation sources, such as those used in medical and dental diagnostic procedures.

One source of natural radiation is cosmic radiation. The earth and all living things on it are constantly being bombarded by radiation from space. The sun and stars emit EM radiation of all wavelengths. Charged particles from the sun and stars interact with the earth's atmosphere and magnetic field to produce a shower of radiation, typically beta and gamma radiation. The dose from cosmic radiation varies in different parts of the world due to differences in elevation and the effects of the earth's magnetic field.

Radioactive material is also found throughout nature. It occurs naturally in soil, water, plants and animals. The major isotopes of concern for terrestrial radiation are uranium and the decay products of uranium, such as thorium, radium, and radon. Low levels of uranium, thorium, and their decay products are found everywhere. Some of these materials are ingested with food and water, while others, such as radon, are inhaled. The dose from terrestrial sources varies in different parts of the world. Locations with higher concentrations of uranium and thorium in their soil have higher dose levels. All people also have radioactive isotopes, such as potassium-40 and carbon-14, inside their bodies. The variation in dose from one person to another is not as great as the variation in dose from cosmic and terrestrial sources.

There are also a number of man-made radiation sources that present some exposure to the public. Some of these sources include tobacco, television sets, smoke detectors, combustible fuels, certain building materials, nuclear fuel for energy production, nuclear weapons, medical and dental X-rays, nuclear medicine, X-ray security systems and industrial radiography. So far, the most significant source of man-made radiation exposure to the average person is from medical procedures, such as diagnostic X-rays, nuclear medicine, and radiation therapy.

2.2 Natural Background Sources

Natural background radiation comes from the following three sources (Figure 2.2):

- Cosmic Radiation
- Terrestrial Radiation
- Internal Radiation

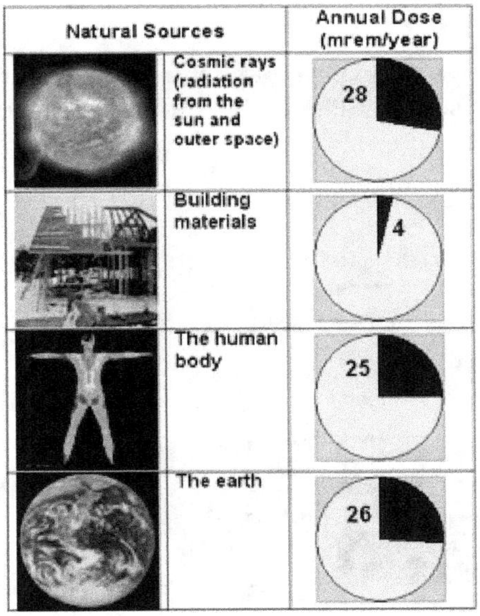

Figure 2.2: Natural Sources.

2.2.1 Cosmic Radiation

The sun and stars send a constant stream of cosmic radiation to Earth, much like a steady drizzle of rain. Differences in elevation, atmospheric conditions, and the Earth's magnetic field can change the amount (or dose) of cosmic radiation that we receive.

2.2.2 Terrestrial Radiation

The Earth itself is a source of terrestrial radiation. Radioactive materials (including uranium, thorium, and radium) exist naturally in soil and rock. Essentially all air contains radon, which is responsible for most of the doses that Americans receive each year from natural background sources. In addition, water contains small amounts of dissolved uranium and thorium, and all organic matter (both plant and animal) contains radioactive carbon and potassium. Some of these materials are ingested with food and water, while others (such as radon) are inhaled. The dose from terrestrial sources varies in different parts of the world, but locations with higher soil concentrations of uranium and thorium generally have higher doses.

2.2.3 Internal Radiation

All people have internal radiation, mainly from radioactive potassium-40 and carbon-14 inside their bodies from birth and, therefore, are sources of exposure to others. The variation in dose from one person to another is not as great as that associated with cosmic and terrestrial sources.

2.3 Man-Made Sources

Although all living things are exposed to natural background radiation, exposure to man-made radiation sources (Figure 2.3) differs for the following groups:

- Members of the Public
- Occupationally Exposed Individuals (Workers)

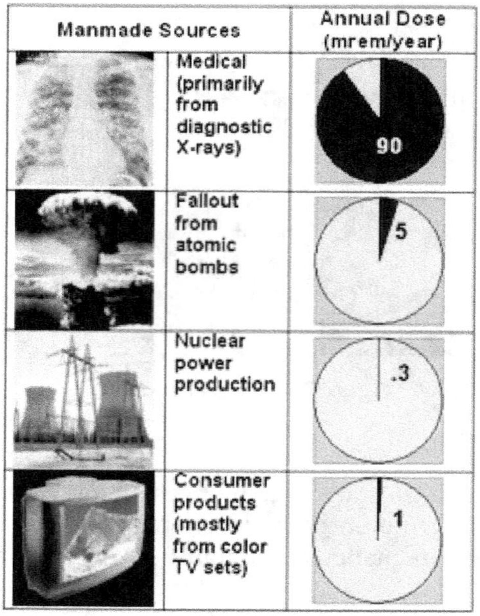

Figure 2.3: Man-made Sources.

2.3.1 Members of the Public

In general, the following man-made sources expose the public to radiation (the significant radioactive isotopes are indicated in parentheses):

Medical Sources (by far, the most significant man-made source)

- Diagnostic X-rays
- Nuclear medicine procedures (iodine-131, cesium-137, and others)
- Consumer Products
- Building and road construction materials
- Combustible fuels, including gas and coal
- X-ray security systems
- Televisions
- Fluorescent lamp starters
- Smoke detectors (americium)
- Luminous watches (tritium)
- Lantern mantles (thorium)
- Tobacco (polonium-210)
- Ophthalmic glass used in eyeglasses
- Some ceramics

To a lesser degree, the public is also exposed to radiation from the nuclear fuel cycle, from uranium mining and milling to disposal of used (spent) fuel. In addition, the public receives some minimal exposure from the transportation of radioactive materials and fallout from nuclear weapons testing and reactor accidents (such as Chernobyl). For that reason, the U.S. Nuclear Regulatory Commission (NRC) requires its licensees to limit the maximum radiation exposure to individual members of the public to 100 mrem (1 mSv) per year.

2.3.2 Occupationally Exposed Individuals

In general, occupationally exposed individuals work in the following areas:

- Fuel cycle facilities
- Industrial radiography
- Radiology departments (medical)
- Nuclear medicine departments
- Radiation oncology departments
- Nuclear power plants
- Government and university research laboratories

Such individuals are exposed to varying amounts of radiation, depending on their specific jobs and the sources with which they work. For that reason, the NRC requires its licensees to limit occupational exposure to adults working with radioactive material to 5,000 mrem (50 mSv) per year. The related NRC regulations and radiation exposure limits are contained in 10 CFR Part 20. Toward that end, employers carefully monitor the exposure of these individuals using instruments called dosimeters.

2.4 Cosmic Rays

Cosmic rays are energetic, subatomic particles that arrive from outside the Earth's atmosphere. The lowest energy cosmic rays are produced by ordinary stars like the Sun. For example, during a solar flare many particles are ejected from the Sun. When these solar particles interact with the Earth's magnetic field, they tend to spiral into the Earth's magnetic poles where they excite the gas in the atmosphere causing it to glow. This is observed as Aurora, or the Northern/Southern Lights (Figure 2.4).

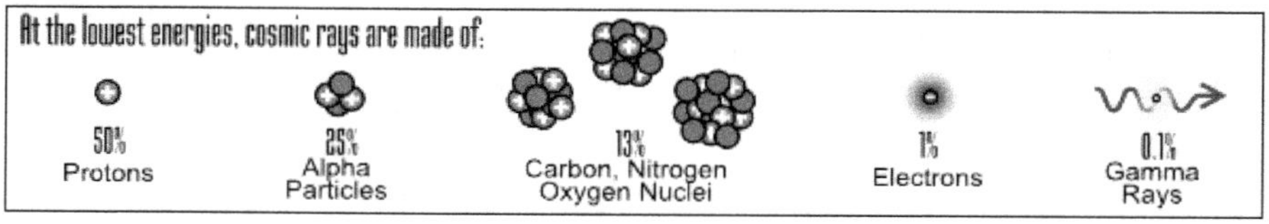

Figure 2.4: Aurora.

Cosmic rays arrive with a variety of energies. At the lowest energies, cosmic radiation is very common. These particles originate from sources close to Earth such as the Sun. We are shielded from most of the low energy particles by their interaction with the Earth's atmosphere and magnetic field.

The basic unit of measurement used in discussing the energy of a cosmic ray is the electron volt, or eV. One eV is the amount of energy gained by a single electron when passing through an electric potential of one volt. Common batteries (such as a D-cell) have a potential of 1.5 volts, so a single electron through a regular battery will gain 1.5 eV (Figure 2.4).

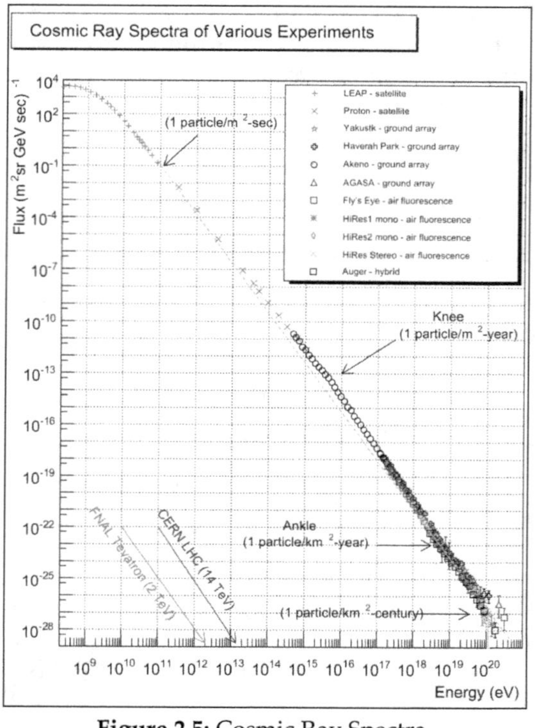

Figure 2.5: Cosmic Ray Spectra.

As the energy of the cosmic rays increases, the frequency of the particles decreases steeply. On

arrival at the Earth, their path experiences less bending in the magnetic field and they penetrate deeper into the atmosphere. Particles with intermediate energy levels, around what is called the "Knee" of the spectrum, are called Very High Energy cosmic rays and they occur with a frequency of one per square meter per year. For the highest energy cosmic rays, above 1016 eV or what is called the "Ankle," the rate of events falls to one per square kilometer per century.

When a highly energetic cosmic ray hits Earth, we would not detect it directly, we would detect something called the air shower of a cosmic ray. Air showers are a cascade of secondary particles created when the cosmic rays smash into the upper atmosphere. As these powerful particles crash through the air, they can spread out over 40 kilometers by the time they reach the ground (and up to billions of secondary particles). The relatively low energy cosmic rays from the sun (aka solar wind) are also responsible for auroras at Earth's poles, and some scientists wonder whether cosmic rays spark lightning (Figure 2.6).

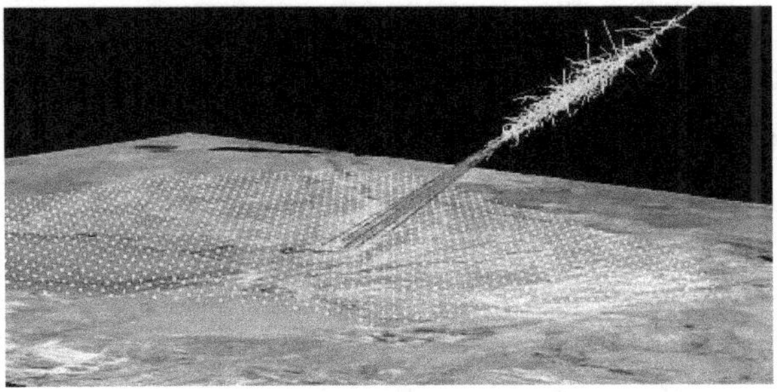

Figure 2.6: Simulation of Cosmic Rays Shower.

Cosmic rays coming from outside the solar system are called galactic cosmic rays, of which many were launched here long ago by distant supernovae explosions. However most of these never reach us. A large percentage are dispersed by the helio sheath of the sun, then another large portion of the remnants are deflected by the Earth's magnetic field, and still some of those left hit the atmosphere but are not powerful enough to reach the ground (Figure 2.7).

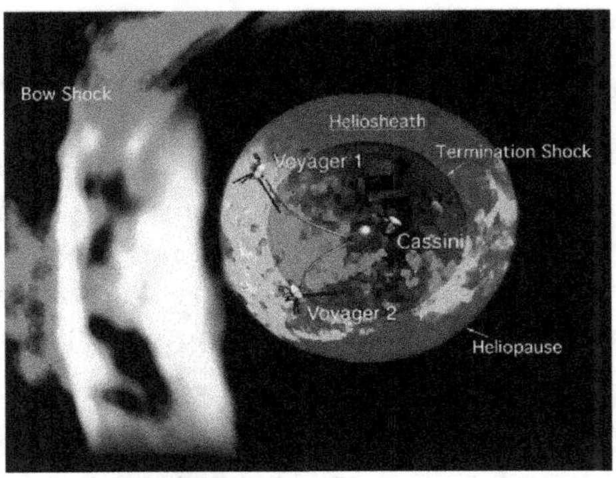

Figure 2.7: Helio Sheath of Sun.

Cosmic rays found beyond Earth's atmosphere are called primary cosmic rays. Sometimes, these primary rays collide with Earth's atmosphere. When this happens, atoms that make up the gas in Earth's atmosphere are smashed apart into fragments which shower down onto Earth's surface. Particles produced from the collision of a primary cosmic ray with Earth's atmosphere are called secondary cosmic rays. These are the guys moving through you right now.

Astronomers think the lower-energy cosmic rays come from within the Milky Way, particularly from star-forming regions and supernova remnants. The more energetic ones we're talking 1,000 trillion eV are thought to come from outside the galaxy. The best way to tell is by determining which direction the cosmic rays come from. But that's not easy. The Milky Way's magnetic field scrambles these particles' paths, analogous to how thick fog scatters light, John Gallagher III and Francis Halzen. "Just as light seems to come from all directions because it scatters over short distances in fog, cosmic rays scattered by magnetic fields also appear to arrive uniformly across the sky," they explain.

2.5 Radioactive Sources

2.5.1 Introduction

Radioactive materials are all around us. Some are beneficial, while others may cause us harm.

Everything on Earth is made of atoms, but you can see that not everything on Earth is the same. This is because everything is made of different types of atoms called elements. Atoms are really small, but each atom of an element is made of even smaller subatomic particles. These are the protons, neutrons, and electrons of the atom.

The number of protons determines what element that atom is. For example, an atom with 8 protons is always oxygen, and an atom with 80 protons is always mercury. If you change the number of protons, you change the element.

One thing that can change though is the number of neutrons. Let's look at carbon for example. A carbon atom will always have 6 protons, but it may have 6, 7, or 8 neutrons. These atoms are carbon isotopes, because they are atoms of the same element that have different numbers of neutrons.

When isotopes are unstable, *i.e.* that they have an imbalance of neutrons and protons, they are radioactive. The carbon atom with 6 neutrons, also called carbon-12 since we just add up the number of protons and neutrons to get its name, and the carbon atom with 7 neutrons, also called carbon-13, are both stable. This is because the number of protons and neutrons is in good balance.

The carbon atom with 8 neutrons is not stable, though. This atom is radioactive because it has too much energy, making it unstable. After a while, the extra energy will be released from the unstable atom. This process is called radioactive decay.

After enough energy is released, the atom returns to a stable state and is no longer radioactive. All radioactive materials eventually decay, just not at the same rate.

Alpha, Beta, & Gamma-rays

The extra energy, or radiation, emitted by radioactive elements comes in three different types: alpha, beta, and gamma. Alpha radiation is a stream of alpha particles, which are positively charged. They're fairly large, which means they have a difficult time getting through materials like clothes and paper.

Beta radiation is a stream of beta particles, which are negatively charged. These particles can more easily penetrate materials like clothes, and they can even get deep into your skin where they can do harm to your cells. But they can be blocked with denser materials like aluminum.

Gamma radiation is high-frequency electromagnetic radiation. Gamma-rays have no charge but have more energy than even visible light or X-rays, so they are able to pass right through most materials, making them quite dangerous. But they can't penetrate very dense materials like lead, which is why you may be given a lead vest for protection in a hospital or laboratory.

2.5.2 Types of Radioactive Source

A radioactive source or radiation source is a sample of a radionuclide, and emits ionizing radiation (one or more of gamma-rays, alpha particles, beta particles, and neutron radiation). Generally, sources are used for irradiation, where the radiation performs a significant ionizing function on a target material, or as a meteorology calibration source, which is used for the calibration of radiometric process and radiation protection instrumentation. They can be sealed in a container or on a surface, or they can be free to move within a fluid.

Many radioactive sources are sealed, meaning they are permanently either completely contained in a capsule or firmly bonded solid to a surface. Capsules are usually made of stainless steel, titanium, platinum or another inert metal. The use of sealed sources removes almost all risk of dispersion of radioactive material into the environment due to mishandling, but the container is not intended to attenuate radiation, so further shielding is required for radiation protection.

Sealed sources are used in almost all applications, where the source does not need to be chemically or physically included in a liquid or gas. Radioactive isotopes emitting ionizing radiation are used in a wide range of applications in the petroleum industry, from exploration and production to platforms to refineries.

Radiation can come from both natural and artificial sources. Radioactive substances are artificially produced during the nuclear fission process. Radiation also comes from substances that are all around us every day. Examples of these natural sources of radiation include soil, sea-water and cosmic rays from space. But radiation effect is measured by its intensity and not whether it is artificial or natural in origin (Figure 2.8).

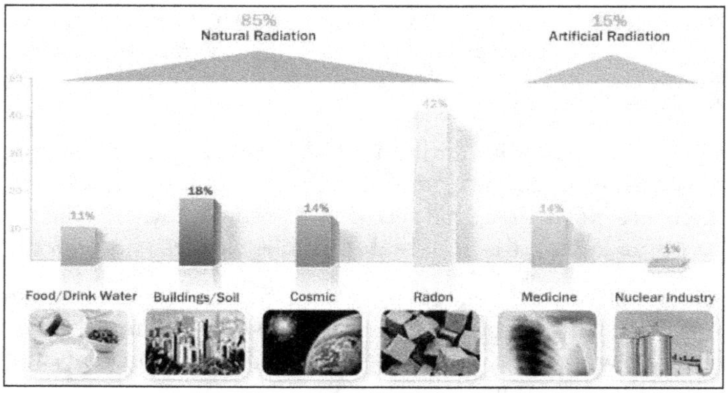

Figure 2.8: About 85% of our Radiation is Completely Natural.

2.6 Category of Radioactive Sources

The categories for radioactive sources are defined by IAEA's Code of Conduct so as to help ensure that radioactive sources are used within an appropriate framework of radiation safety and security.

2.6.1 Category 1 Sources

If not safely managed or securely protected, would be likely to cause permanent injury to a person who handled them or was otherwise in contact with them for more than a few minutes. It would probably be fatal to be close to this amount of unshielded material for a period of a few minutes to an hour. These sources are typically used in practices such as radio-thermal generators, irradiators, and radiation tele-therapy.

2.6.2 Category 2 Sources

If not safely managed or securely protected, could cause permanent injury to a person who handled them or was otherwise in contact with them for a short time (minutes to hours). It could possibly be fatal to be close to this amount of unshielded radioactive material for a period of hours to days. These sources are typically used in practices such as industrial gamma radiography, high dose rate brachytherapy and medium dose rate brachytherapy.

2.6.3 Category 3 Sources

If not safely managed or securely protected, could cause permanent injury to a person who handled them or was otherwise in contact with them for some hours. It could possibly -- although it is unlikely -- be fatal to be close to this amount of unshielded radioactive material for a period of days to weeks. These sources are typically used in practices such as fixed industrial gauges involving high activity sources (for example, level gauges, dredger gauges, conveyor gauges, and spinning pipe gauges) and well logging.

Description	Activity	Half-Life	Item#
Barium-133	1, 5 & 10uCi	10.7 years	BA-133-X
	Primary Radiation Type	Energies KeV	Secondary Radiation Type(s)
	Gamma (γ)	81.0, 276.3, 303.7, 355.9, 383.7	N/A

Description	Activity	Half-Life	Item#
Cesium-137	0.25, 0.5, 1, 5 & 10uCi	30.1 years	CS-137-X
	Primary Radiation Type	Energies KeV	Secondary Radiation Type(s)
	Gamma (γ) & Beta (β)	γ: 32, 661.6 β: 511.6, 1173.2	Multiple Radiation Types: (γ) & (β)

Description	Activity	Half-Life	Item#
Cadmium-109	1, 5 & 10uCi	465 days	CD-109-S-X
	Primary Radiation Type	Energies KeV	Secondary Radiation Type(s)
	Gamma (γ)	88.0	N/A

Description	Activity	Half-Life	Item#
Cobalt-57	1, 5, 10, 25, 50 & 100uCi	271 days	CO-57-X
	Primary Radiation Type	Energies KeV	Secondary Radiation Type(s)
	Gamma (γ)	122.1, 136.4	Beta (+β) 19 keV

Description	Activity	Half-Life	Item#
Cobalt-60	1uCi	5.27 years	CO-60-1
	Primary Radiation Type	Energies KeV	Secondary Radiation Type(s)
	Gamma (γ)	γ: 1173.2, 1332.5 β: 317.9	Beta (β) 317.9

Description	Activity	Half-Life	Item#
Europium 152	1uCi	13.516 years	Eu-152-1
	Primary Radiation Type	Energies KeV	Secondary Radiation Type(s)
	Gamma (γ) & Beta (β)		N/A

Description	Activity	Half-Life	Item#
Manganese-54	1, 5 & 10uCi	312 days	MN-54-X
	Primary Radiation Type	Energies KeV	Secondary Radiation Type(s)
	Gamma (γ) & Beta (β)	γ: 834.8 β: 542.2	Beta (β) 542.2

Description	Activity	Half-Life	Item#
Sodium-22	1, 5, & 10uCi	2.6 years	NA-22-X
	Primary Radiation Type	Energies KeV	Secondary Radiation Type(s)
	Gamma (γ)	511.0, 1274.5	Beta (+β) 546 keV
Description	**Activity**	**Half-Life**	**Item#**
Polonium-210	0.05, & 0.1uCi	138 days	PO-210-X
	Primary Radiation Type	Energies KeV	Secondary Radiation Type(s)
	Alpha (α)	5304.5	N/A
Description	**Activity**	**Half-Life**	**Item#**
Tin-113	1uCi	115 days	SN-113-1
	Primary Radiation Type	Energies KeV	Secondary Radiation Type(s)
	Beta (β)	14.38	Gamma (γ)
Description	**Activity**	**Half-Life**	**Item#**
Strontium-90	0.1uCi	28.5 years	SR-90-0.1
	Primary Radiation Type	Energies KeV	Secondary Radiation Type(s)
	Beta (β)	546	N/A
Description	**Activity**	**Half-Life**	**Item#**
Thallium-204	1, 5, &10uCi	3.78 years	TL-204-X
	Primary Radiation Type	Energies KeV	Secondary Radiation Type(s)
	Beta (β-)	763.7	N/A
Description	**Activity**	**Half-Life**	**Item#**
Zinc-65	1uCi	244 days	ZN-65-1
	Primary Radiation Type	Energies KeV	Secondary Radiation Type(s)
	Gamma (γ) & Beta (β)	γ: 1115.6 β: 236.34, 1351.9	Beta (β)

Minute quantities of radioactive atoms are used to meet the needs of medicine, industry or research, as radioactive sources or tracers. These radio-nuclides that do not exist for most of them in a natural state are generally produced in specialized installations small reactors designed for that purpose or, in the case of some very short-lived isotopes used in medicine in large hospitals, small accelerators installed permanently.

These radio-nuclides can cause an accidental exposure to radioactivity if precautions are not taken. Dispersal of radio-nuclides when they present a danger because of their activity, the noxiousness of rays or the life of the radioisotope must be avoided. That's why the radioactive sources are usually sealed to prevent the dispersal of radioactive material. When they are not used, they are stored in a container or a "castle" in order to avoid unnecessary exposure to radiation.

The main radio-protection problem of radiation with radioactive sources is their large numbers, a consequence of their multiple applications. Thus, the Nuclear Safety Authority must manage some 50,000 radioactive sources used in industry, research and medicine. Another problem is the end of life of these sources, which become waste managed by the National Agency for Radioactive Waste.

The tracers can be considered as radioactive sources that are not sealed. The quantities of radio-nuclides in question are much smaller and shorter lifetimes, allowing their dispersal. They are widely used in medical diagnosis.

The use of radioactive tracers incorporated into molecules is very common in biological research. They are thus a powerful tool of investigation in cell and molecular biology. The main radionuclide used as tracers are phosphorus 32 or 33, carbon 14, sulfur 35, chromium 51, iodine 125 and tritium. They are used as tracers and for calibration or instruction purposes.

A plethora of radioactive tracers are used in industrial processes to investigate the causes of problems or losses of performance, measure flow, controlling mixtures of materials, locate leaks in heat exchangers and piping. For example, the oil industry uses radioisotopes to identify leaks in its oil and gas pipelines.

Source Capsule for Teletherapy Unit

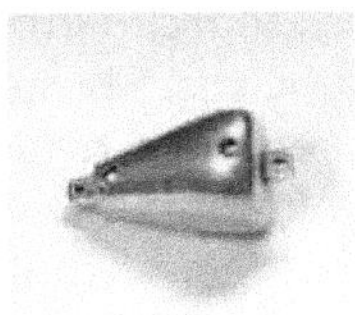

Radium Industrial Radiography Source

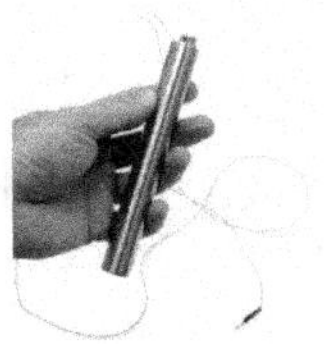

Californium-252 Sources

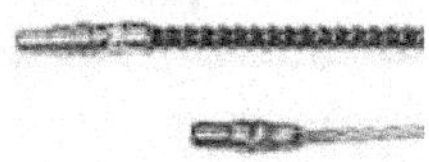

Industrial Radiography Sources

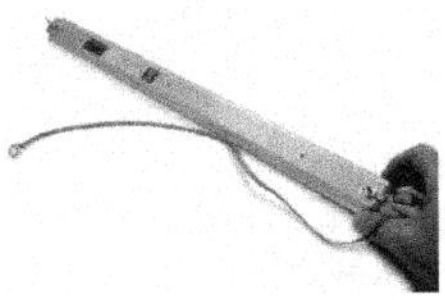

NRD "Nuclestat" Static Eliminator

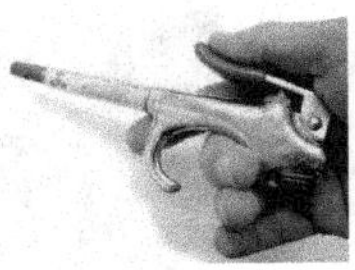

NRD "Ion Gun" Static Eliminator

NRD "Ion Gun" Static Eliminator

3M Line Source

ORNL Annular Source

ORNL Rectangular Bar Source

ORNL Brass Sources

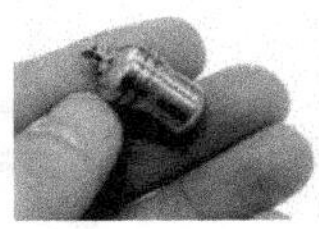

Early Industrial Radiography Source

AECL Co-60 Source

Small Annular Source

Cylindrical Source

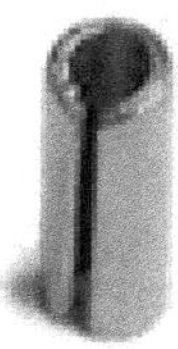

Small Slotted Annular Source

2.6.4 Unknown Sources

Brach therapy Capsule

Mystery Spherical Source

Dome-shaped Source

Mystery Cylindrical Sources

2.7 Accelerators

Several new techniques have been demonstrated that can produce large-amplitude, coherent, high-phase-velocity electron plasma waves. These include the beat wave and wake field concepts. An electron beam accelerated by a beat wave accelerator is pictured in Figure 2.9. Such beat wave accelerators have achieved accelerated electron energies of 9 MeV within distances of 1.5 cm. Attaining 500-MeV energy gains at GeV-per-meter rates is considered a plausible goal within a five-year period in which progress, funding priority, and follow-on application potential may be assessed.

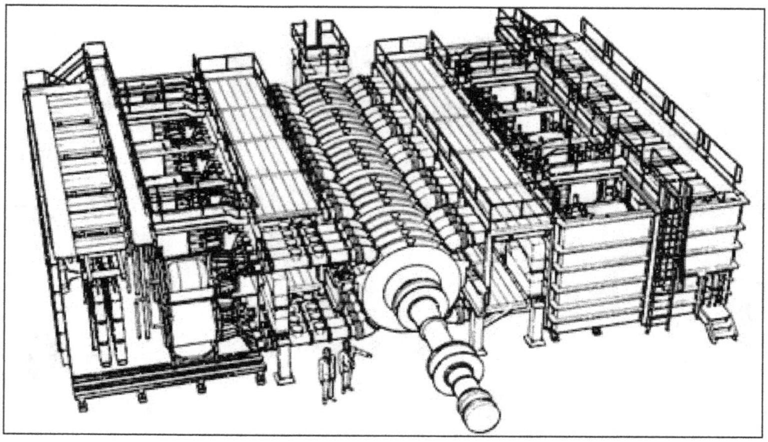

Figure 2.9: Beat Wave Accelerator.

Particles were accelerated in electric fields through an evacuated region of space, such as the interior of a vacuum tube. A beam of particles passing through air would quickly get destroyed through scattering and absorption, so that vacuum technology necessarily preceded the development of accelerators.

The natural unit of energy of an elementary particle, such as an electron or a proton, is the electron Volt (eV), the energy given to a particle of unit charge (e =1.6 ′10^{-19} **Coulombs**) in traveling through a potential difference of one Volt: 1 eV = 1.6 ′ 10^{-19} **Joules**. The energies available in electron and proton accelerators have steadily increased over the years, and today machines are being operated at energies of about 1 Tera electron Volt (TeV) = 10^{12} **eV**.

The use of particle accelerators to produce, identify, and study the elementary constituents of matter has revolutionized our conception of matter. Particle accelerators have also stimulated a wide variety of technological applications throughout the twentieth century, which continue at an increasing pace today.

2.7.1 Types

1. Particle Accelerator

It is a machine that accelerates elementary particles, such as electrons or protons, to very high energies. On a basic level, particle accelerators produce beams of charged particles that can be used for a variety of research purposes. There are two basic types of particle accelerators:

- Linear accelerators
- Circular accelerators

Particle accelerators use electric fields to speed up and increase the energy of a beam of particles, which are steered and focused by magnetic fields. The particle source provides the particles, such as protons or electrons that are to be accelerated. The beam of particles travel inside a vacuum in the metal beam pipe. The vacuum is crucial to maintaining an air and dust free environment for the beam of particles to travel unobstructed. Electromagnets steer and focus the beam of particles while it travels through the vacuum tube.

Electric fields spaced around the accelerator switch from positive to negative at a given frequency, creating radio waves that accelerate particles in bunches. Particles can be directed at a fixed target, such as a thin piece of metal foil, or two beams of particles can be collided. Particle detectors record and reveal the particles and radiation that are produced by the collision between a beam of particles and the target.

Particle accelerators are essential tools of discovery for particle and nuclear physics and for sciences that use X-rays and neutrons, a type of neutral subatomic particle. Particle physics, also called high-energy physics, asks basic questions about the universe. With particle accelerators as their primary scientific tools, particle physicists have achieved a profound understanding of the fundamental particles and physical laws that govern matter, energy, space and time (Figure 2.10).

Tens of millions of patients receive accelerator-based diagnoses and therapy each year in hospitals and clinics around the world. There are two primary roles for particle accelerators in medical applications: the production of radioisotopes for medical diagnosis and therapy, and as sources of beams of electrons, protons and heavier charged particles for medical treatment.

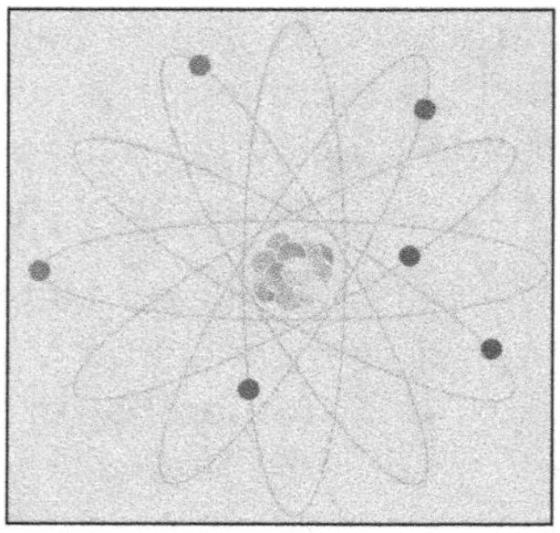

Figure 2.10: Particle Accelerator.

The wide range of half-lives of radioisotopes and their differing radiation types allow optimization for specific applications. Isotopes emitting X-rays, gamma-rays or positrons can serve as diagnostic probes, with instruments located outside the patient to image radiation distribution and thus the biological structures and fluid motion or constriction (blood flow, for example). Emitters of beta rays (electrons) and alpha particles (helium nuclei) deposit most of their energy close to the site of the emitting nucleus and serve as therapeutic agents to destroy cancerous tissues.

Radiation therapy by external beams has developed into a highly effective method for treating cancer patients. The vast majority of these irradiations are now performed with microwave linear accelerators producing electron beams and X-rays. Accelerator technology, diagnostics and treatment technique developments over the past 50 years have dramatically improved clinical outcomes. Today, 30 proton and three carbon-ion-beam treatment centers are in operation worldwide, with many new centers on the way.

Linear Particle Accelerator

A linear particle accelerator is a type of particle accelerator that greatly increases the kinetic energy of charged subatomic particles or ions by subjecting the charged particles to a series of oscillating electric potentials along a linear beam line. The principles for such machines were proposed by Gustav Ising in 1924, while the first machine that worked was constructed by Rolf Widerøe in 1928 at the RWTH Aachen University. Linacs have many applications: they generate X-rays and high energy electrons for medicinal purposes in radiation therapy, serve as particle injectors for higher-energy accelerators, and are used directly to achieve the highest kinetic energy for light particles (electrons and positrons) for particle physics. Linear particle accelerator comprises of shown in Figure 2.11.

Particle Source

The design of the source depends on the particle that is being moved. Electrons are generated by a cold cathode, a hot cathode, a photo-cathode, or radio-frequency (RF) ion sources. Protons are generated in an ion source, which can have many different designs. If heavier particles are to be accelerated, (*e.g.*, uranium ions), a specialized ion source is needed. A high voltage source for the initial injection of particles.

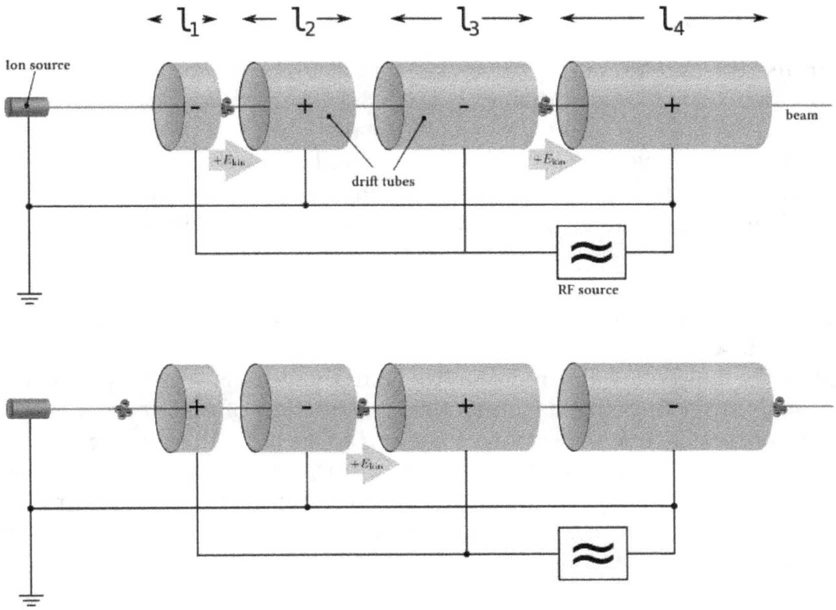

Figure 2.11: Parts of Linear Particle Accelerator.

Hollow Pipe Vacuum C

The length will vary with the application. If the device is used for the production of X-rays for inspection or therapy the pipe may be only 0.5 to 1.5 meters long. If the device is to be an injector for a synchrotron it may be about ten meters long. If the device is used as the primary accelerator for nuclear particle investigations, it may be several thousand meters long. Within the chamber, electrically isolated cylindrical electrodes are placed, whose length varies with the distance along the pipe. The length of each electrode is determined by the frequency and power of the driving power source and the nature of the particle to be accelerated, with shorter segments near the source and longer segments near the target.

The mass of the particle has a large effect on the length of the cylindrical electrodes; for example an electron is considerably lighter than a proton and thus will generally require a much smaller section of cylindrical electrodes as it accelerates very quickly. Likewise, because its mass is so small, electrons have much less kinetic energy than protons at the same speed. Because of the possibility of electron emissions from highly charged surfaces, the voltages used in the accelerator have an upper limit, so this can't be as simple as just increasing voltage to match increased mass. One or more sources of radio-frequency energy, used to energize the cylindrical electrodes. A very high power accelerator will use one source for each electrode. The sources must operate at precise power, frequency and phase appropriate to the particle type, to be accelerated to obtain maximum device power.

Appropriate Target

If electrons are accelerated to produce X-rays, then a water cooled tungsten target is used. Various target materials are used when protons or other nuclei are accelerated, depending upon the specific investigation. For particle-to-particle collision investigations the beam may be directed to a pair of storage rings, with the particles kept within the ring by magnetic fields. The beams may then be extracted from the storage rings to create head on particle collisions.

Advantages

Linacs of appropriate design are capable of accelerating heavy ions to energies exceeding those available in ring-type accelerators, which are limited by the strength of the magnetic fields required to maintain the ions on a curved path. High power linacs are also being developed for production of electrons at relativistic speeds, required since fast electrons traveling in an arc will lose energy through synchrotron radiation; this limits the maximum power that can be imparted to electrons in a synchrotron of given size. Linacs are also capable of prodigious output, producing a nearly continuous stream of particles, whereas, a synchrotron will only periodically raise the particles upto sufficient energy to merit a "shot" at the target.

The high density of the output makes the linac particularly attractive for use in loading storage ring facilities with particles in preparation for particle to particle collisions. The high mass output also makes the device practical for the production of antimatter particles, which are generally difficult to obtain, being only a small fraction of a target's collision products. These may then be stored and further used to study matter-antimatter annihilation.

Linac-based radiation therapy for cancer therapy began with treatment of the first patient in 1953 in London at Hammersmith Hospital, with an 8 MV machine built by Metropolitan-Vickers, as the first dedicated medical linac. A short while later in 1955, 6 MV linac therapy from a different machine was being used in the United States.

Medical grade linacs accelerate electrons using a tuned-cavity waveguide, in which the RF power creates a standing wave. Some linacs have short, vertically mounted waveguides, while higher energy machines tend to have a horizontal, longer waveguide and a bending magnet to turn the beam vertically towards the patient. Medical linacs use mono energetic electron beams between 4 and 25 MeV, giving an X-ray output with a spectrum of energies up to 25 MeV and including the electron energy when the electrons are directed at a high-density (such as tungsten) target. The electrons or X-rays can be used to treat both benign and malignant disease. The LINAC produces a reliable, flexible and accurate radiation beam. The versatility of LINAC is a potential advantage over cobalt therapy as a treatment tool. In addition, the device can simply be powered off when not in use; there is no source requiring heavy shielding – although the treatment room itself requires considerable shielding of the walls, doors, ceiling *etc.* to prevent escape of scattered radiation. Prolonged use of high powered (>18 MeV) machines can induce a significant amount of radiation within the metal parts of the head of the machine after power to the machine has been removed.

2. The Cyclotron

In 1928 Ernest Lawrence of the University of California, inspired by the work of Widerøe, had the idea of utilizing a curved path for a particle accelerator. A magnetic field perpendicular to the plane of motion of an accelerated particle will result in the particle taking a curved path. By studying the simple relationship between the forces acting on the particle, Lawrence realised that the increase in the radius of the path taken by the particle is compensated for, by the increased velocity of the particle if the magnetic field, the charge of the particle and the particles' mass remain constant. With this in mind, he built what became known as a Cyclotron (Figure 2.12).

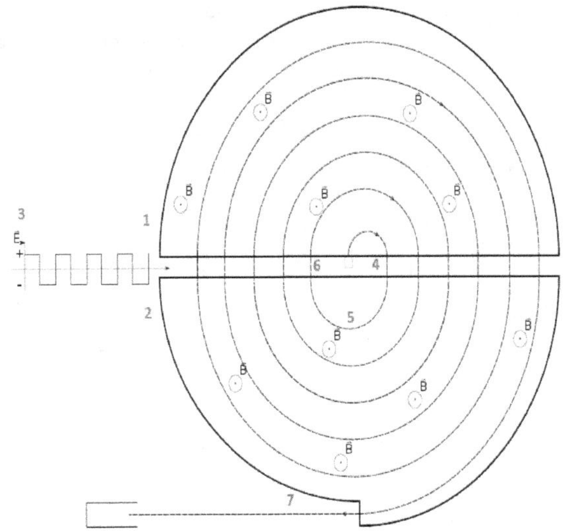

Figure 2.12: Basic Diagram of a Cyclotron.

Cyclotron consist of two hollow D-shaped electrodes (1 & 2) alternatively charged to a voltage by an oscillator (3). The electrodes were separated by a small gap. When one of the electrodes is charged, a particle is accelerated across the gap into the other (4), where, under the influence of a magnetic field, it moves in a semi-circular path back to the surface of the electrode (5). Just as the voltage has charged the other electrode, the particle is again accelerated across the gap (6). As the speed of the particle increases, radius of semi-circular motion of particle increases till the particles are eventually focused out of Cyclotron as a high energy beam (7).

Using this design, Lawrence's first accelerator, built in 1931, was able to accelerate beams of Hydrogen ions to 80keV. After some modifications to the original design, Lawrence was able to ramp up the energy of the output beam to 1.22MeV (1,220,000eV) and repeat Cockcroft & Walton's result of nuclear disintegration of Lithium by 1932. The Cyclotron became the accelerator of choice across the world from many years as it was more cost and size effective than other accelerators.

Synchrocyclotron

The Synchrocyclotron is a modification of the classic Cyclotron design developed to overcome the relativistic limitations on the classic Cyclotron as described in the previous section. The Synchrocyclotron differs from the classic Cyclotron in that the frequency of the AC voltage does not remain constant, it synchronizes with the orbit frequency of the accelerated particles. Put simply, as the particle energy, and therefore mass, increases, the time it takes for the particle to complete one orbit, increases, *i.e.* it arrives at the surface of the electrode, where it is accelerated, less frequently. If the AC voltage also peaks less frequently so as to match the decreased orbit frequency of the particles being accelerated then the particle and voltage avoid dropping out of sync and the particles can be accelerated beyond the relativistic limit of the classical Cyclotron. The first Synchrocyclotron was a modified Cyclotron completed by researchers at Berkeley. They found that much lower voltages were needed to accelerate particles than in a Cyclotron and by 1947 they were able to accelerate Deuteron (an isotope of Hydrogen) ions to 190MeV and He^{2+} (an isotope of Helium) ions to 380MeV (Figure 2.13).

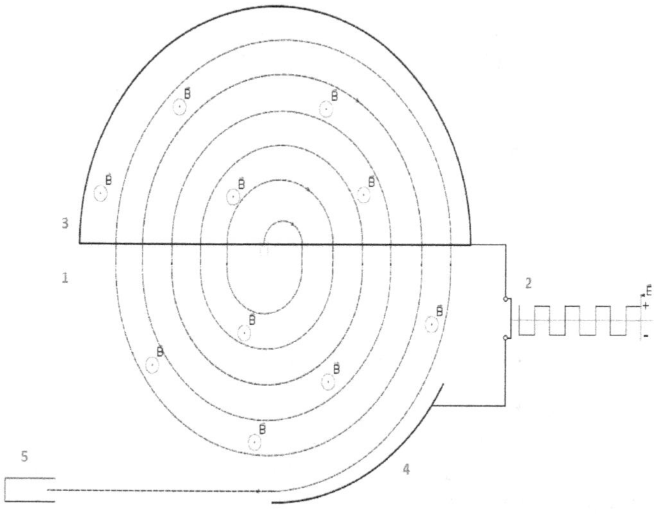

Figure 2.13: Synchrocyclotron.

The modified Cyclotron had one of the D-shaped electrodes removed (1). The particle is accelerated in the same way as in a Cyclotron except that the RF generator responsible for producing the AC voltage which accelerates the particle, is replaced with a variable frequency RF generator (2) so that the AC frequency can be synchronized with orbital frequency of the particle. The AC voltage is applied across the remaining D-shaped electrode (3) and a new deflecting electrode (4) which is responsible for directing the particle out of the accelerator and towards a target.

Synchrotron

The basic principle of the Synchrotron is to maintain the accelerated particles at a constant orbital radius. This is achieved by synchronizing the magnetic field strength with the energy of the accelerated particles. So, as the particles accelerate and gain energy, the magnetic field is increases, keeping the particles orbit constant. The first Synchrotron to be built was a modified Betatron and was completed by two English physicists, Frank Goward and D. Barnes. Many Synchrotrons were built after this and by 1954. Under the leadership of Phillip Dee, a Synchrotron capable of accelerating electrons to 350MeV was built at the University of Glasgow.

The upper limit to the speed of any object is the speed of light, nothing can travel faster. Light particles such as electrons begin to approach the speed of light at fairly low energies. Heavier particles like protons approach the speed of light at much higher energies (a proton is almost 2000 times heavier), so to produce higher energy beams, scientists looked to accelerate protons rather than electrons. Due to the larger mass of protons, accelerating protons in a Synchrotron is slightly more complicated than accelerating electrons. As the protons are accelerated, their orbital frequency increases, meaning that the frequency of the AC Voltage across the accelerator gaps also has to increase. However, the strength of the magnetic field also has to increase to maintain a constant orbital radius, so the orbital frequency of the protons, the frequency of the AC voltage and the increase in magnetic field all have to be synchronised. The first proton Synchrotron to be completed was the 'Cosmotron' at Brookhaven, finished in 1952 and by 1954 it was accelerating protons to 3GeV (3,000,000,000eV).

In 1947 whilst working on their electron Synchrotron, scientists working at General Electric's Research Labs observed for the first time what came to be known as synchrotron radiation. Synchrotron radiation is an electromagnetic radiation, usually in the X-ray range, emitted by charged particles traveling at relativistic speeds, forced to take a curved path by a magnetic field. At first synchrotron radiation was seen as an inconvenience as it caused a loss in energy for the accelerated particles, however it wasn't long until its usefulness in spectroscopy and crystallography was realized and today Synchrotrons are built for the sole purpose of generating synchrotron radiation.

The development of the Synchrotron led to the construction of large scale accelerators, as in a Synchrotron the components can be divided in different sections. The acceleration of the particles and the bending of the particle's path can be split, such that the accelerator, as opposed to being circular, can be slightly oval shaped, or like a square with rounded corners. A particle in a Synchrotron will be accelerated in accelerator gaps whilst traveling along the straight section, as they would be in a Linac. The particle's path is then curved by a magnetic field, taking the particle to another straight section where it is again accelerated. The particle will go round and round the accelerator, gaining more and more energy with every revolution. Below is a Figure 2.14 of a simple proton Synchrotron.

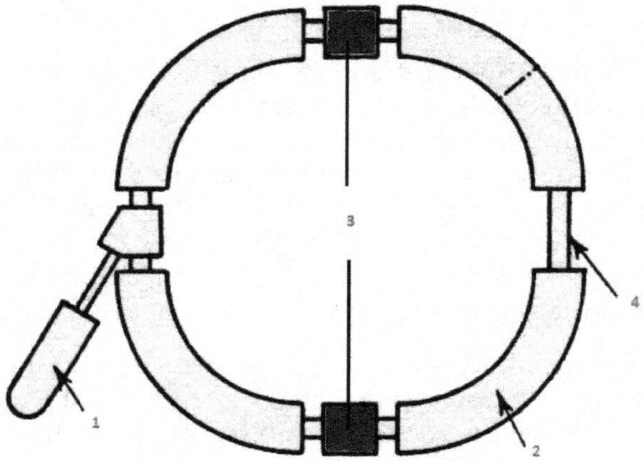

Figure 2.14: Proton Synchrotron Accelerator.

First protons are accelerated in a pre-accelerator, which could be a Cockcroft-Walton or Van de Graaff accelerator, a Linac, or in the case of some larger Synchrotrons like the SPS at CERN even a smaller Synchrotron.

The protons then enter the Synchrotron via an injector. The protons are then made to follow a curved path by bending magnets, which can be either electromagnets or superconducting electromagnets. After leaving the bending magnets the protons travel to an accelerating cavity where they are further accelerated. The particles are again made to follow a curved path bringing them to a second straight section where they can either be accelerated again in another accelerator cavity, or in the case of the example below, simply follow a straight path before being directed again to another bending magnet which will direct the protons to the second accelerator cavity parallel to the first.

The synchrotron is an extremely important type of accelerator which is used today to accelerate electrons, protons and atomic nuclei, in many cutting edge research facilities world-wide.

2.8 Synchrotron Radiation

2.8.1 Introduction

A relativistic electron passing through a bending magnet emits electromagnetic radiation in the forward direction with an extremely narrow opening angle.

The photon density of this 'so-called synchrotron' radiation is correspondingly high and has, in addition, a very broad radiation spectrum. Because of its outstanding properties, synchrotron radiation has become a very powerful tool in basic research and technical applications. The required relativistic electron beam is provided by utilizing modern particle accelerator techniques. Presently the most successful accelerator type used as a dedicated synchrotron radiation source is the storage ring. Very important for a high quality of radiation is strong focusing of the electron beam circulating in the accelerator. The theory of this low-emittance optics is presented, including some important examples of magnet lattices. Inserting special wiggler and undulated magnets into a storage ring gives a significant increase of the photon density.

Synchrotron radiation refers to a continuous band of electromagnetic spectrum including infrared, visible light, ultraviolet, and X-rays. This light has been called "synchrotron radiation", since it was accidentally discovered in an electron synchrotron of the General Electric Company, USA, in 1947.

Synchrotron Radiation - A Form of Light

Light has always been indispensable to man's exploration of nature. All wavelengths of the electromagnetic spectrum can be referred to as light. Light of different wavelengths is used for different purposes. The longest wavelengths, radio waves, are used for observing the expansive universe, and microwaves are used to detect planes, ships, and typhoons. Infrared light is an ideal light source for night vision systems and the detection of missiles through tracking their heat sources. Visible light is the only wavelength humans can see with their naked eyes. Ultra-violet light is used to examine the structure of gas molecules and condensed matter. X-rays are the best source for researching crystal structures; and gamma-rays, with the shortest wavelength, allow researchers to explore the inner world of atoms (Figure 2.15).

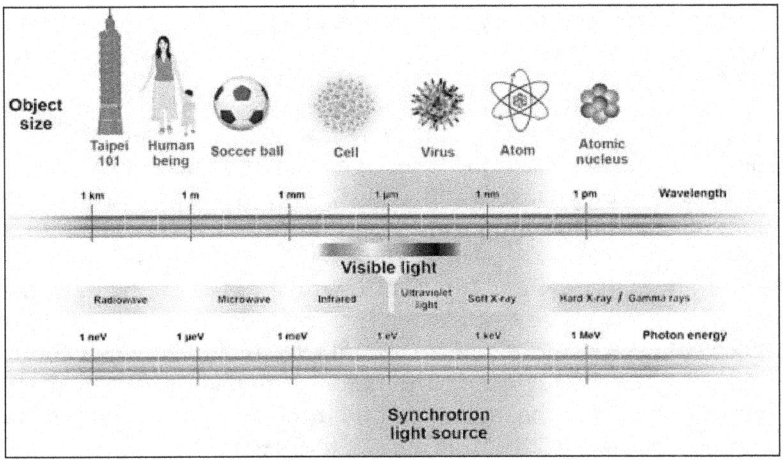

Figure 2.15: Synchrotron Light Source.

2.8.2 Occurrence

Synchrotron radiation may occur in accelerators either as a nuisance, causing undesired energy loss in particle physics contexts, or as a deliberately produced radiation source for numerous laboratory applications. Electrons are accelerated to high speeds in several stages to achieve a final energy that is typically in the giga electron volt range. The electrons are forced to travel in a closed path by strong magnetic fields. This is similar to a radio antenna, but with the difference that the relativistic speed changes the observed frequency due to the Doppler Effect by a factor γ. Relativistic Lorentz contraction bumps the frequency by another factor of γ, thus multiplying the gigahertz frequency of the resonant cavity that accelerates the electrons into the X-ray range. Another dramatic effect of relativity is that the radiation pattern is distorted from the isotropic dipole pattern expected from non-relativistic theory into an extremely forward-pointing cone of radiation. This makes synchrotron radiation sources the most brilliant known sources of X-rays. The planar acceleration geometry makes the radiation linearly polarized when observed in the orbital plane, and circularly polarized when observed at a small angle to that plane.

The advantages of using synchrotron radiation for spectroscopy and diffraction have been realized by an ever-growing scientific community, beginning in the 1960s and 1970s. In the beginning, accelerators were built for particle physics, and synchrotron radiation was used in "parasitic mode" when bending magnet radiation had to be extracted by drilling extra holes in the beam pipes. The first storage ring commissioned as a synchrotron light source was Tantalus, at the Synchrotron Radiation Center, first operational in 1968. As accelerator synchrotron radiation became more intense and its applications more promising, devices that enhanced the intensity of synchrotron radiation were built into existing rings. Third-generation synchrotron radiation sources were conceived and optimized from the outset to produce brilliant X-rays. Fourth-generation sources that will include different concepts for producing ultra-brilliant, pulsed time-structured X-rays for extremely demanding and also probably yet-to-be-conceived experiments are under consideration.

Bending electromagnets in accelerators were first used to generate this radiation, but to generate stronger radiation, other specialized devices – insertion devices – are sometimes employed. Current (third-generation) synchrotron radiation sources are typically reliant upon these insertion devices, where straight sections of the storage ring incorporate periodic magnetic structures with many magnets in a pattern of alternating N and S poles that force the electrons in sinusoidal or helical path. Thus, instead of a single bend, many tens or hundreds of "wiggles" at precisely calculated positions add up or multiply the total intensity of the beam. These devices are called wigglers or undulators. The main difference between an undulator and a wiggler is the intensity of their magnetic field and the amplitude of the deviation from the straight line path of the electrons.

2.8.3 Principle of Synchrotron Radiation

The basic principle behind the generation of synchrotron radiation was discovered by German physicist Heinrich Hertz. He found that charged particles emit electromagnetic radiation when accelerated. This is because the necessary rearrangement of the electric field around the charge causes a perturbation to radiate outwards at the speed of light: the electromagnetic radiation. This principle is used in for example, radio or TV antennae, where charges move up and down periodically in the metal at a frequency specific to the station or channel, thus generating the signal.

The same principle applies on free electrons (or positrons) in an accelerator, although now the source of the radiation is moving. In this case the radiation is seen in the laboratory frame as concentrated in the direction of the electrons' movement. The effect can be compared to objects being thrown off a fast moving vehicle which would all fly more or less in the same direction. Electrons in an accelerator move almost at the speed of light, *i.e.* they are "relativistic". For these electrons, the effect is further magnified by relativity where beam is concentrated into a forward cone with half angle of typically 0.1 to 1 mrad (1 mrad ~ 0.05 degrees) depending on the electrons' energy which can be measured by the Lorentz factor γ. It relates rest energy and actual energy of the electron through $\gamma = E/E0$, where E0=0.511MeV is the electron's rest energy.

2.8.4 Properties

- High Intensity
- Continuous Spectrum
- Excellent Collimation
- Low Emittance
- Pulsed-time Structure
- Polarization

As an example, the intensity of synchrotron X-rays is more than a million times higher than that of X-rays from a conventional X-ray tube. Experiments that took a month to complete can now be done in only a few minutes. With synchrotron radiation, molecular structures that once baffled researchers can now be analyzed precisely, and this progress has opened up many new research fields over the last few years (Figure 2.16).

2.9 Neutron Sources

2.9.1 Introduction

Neutron source appears as a device that emits neutrons, irrespective of the mechanism used to produce the neutrons. Neutron sources are used in physics, engineering, medicine, nuclear weapons, petroleum exploration, biology, chemistry, and nuclear power. Neutron source variables include the energy of the neutrons emitted by the source, the rate of neutrons emitted by the source, the size of the source, the cost of owning and maintaining the source, and government regulations related to the source.

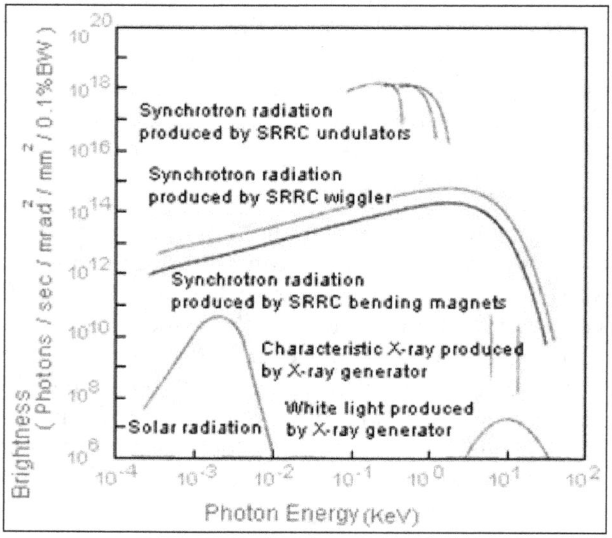

Figure 2.16: Progress Curves.

Neutrons are produced when alpha particles impinge upon any of several low-atomic-weight isotopes including isotopes of beryllium, carbon, and oxygen. This nuclear reaction can be used to construct a neutron source by mixing a radioisotope that emits alpha particles such as radium, polonium, or americium with a low-atomic-weight isotope, usually by blending powders of the two materials. Typical emission rates for alpha reaction, neutron sources range from 1×106 to 1×10^8 neutrons per second.

As an example, a representative alpha-beryllium neutron source can be expected to produce approximately 30 neutrons for every one million alpha particles. The useful lifetime for these types of sources is highly variable, depending upon the half-life of the radioisotope that emits the alpha particles. The size and cost of these neutron sources are comparable to spontaneous fission sources. Usual combinations of materials are plutonium-beryllium (PuBe), americium-beryllium (AmBe), or americium-lithium (AmLi).

Neutrons are produced when photons above the nuclear binding energy of a substance are incident on that substance, causing it to undergo giant dipole resonance after which it either emits a neutron (photoneutron) or undergoes fission (photofission). The number of neutrons released by each fission event is dependent on the substance. Typically photons begin to produce neutrons on interaction with normal matter at energies of about 7 to 40 MeV, which means that radiotherapy facilities using megavoltage X-rays also produce neutrons, and some require neutron shielding. In addition, electrons of energy over about 50 MeV may induce giant dipole resonance in nuclides by a mechanism which is the inverse of internal conversion, and thus produce neutrons by a mechanism similar to that of photoneutrons.

2.9.2 Sources

In addition to neutron-induced fission, neutrons are produced by other reactions. The neutrons produced by reactions other than neutron-induced fission are called source neutrons. Source neutrons are important because they ensure that the neutron population remains high enough to allow a visible indication of neutron level on the most sensitive monitoring instruments while the reactor is shutdown during the startup sequence. This verifies instrument operability and allows monitoring of neutron population changes. Source neutrons can be classified as either intrinsic or installed neutron sources.

2.9.3 Types of n-Sources

Small and portable sources (small n yield):

- Natural sources (spontaneous fission)
- Small radioactive sources: (α,n) reactions
- Electrostatic accelerating machines: (D-T) reactions

Big and expensive facilities (large n yield):

- Research Reactors
- Spallation sources

Large (Significant) Neutron Sources

Nuclear Reactors

There are nuclei that can undergo fission on their own spontaneously, but only certain nuclei, like uranium-235, uranium-233 and plutonium-239, can sustain a fission chain reaction. This is because these nuclei release neutrons when they break apart, and these neutrons can induce fission of other nuclei. Uranium-235 which exists as 0.7% of naturally occurring uranium undergoes nuclear fission with thermal neutrons with the production of, on average, 2.4 fast neutrons and the release of ~ 180 MeV of energy per fission. Free neutrons released by each fission play very important role as a trigger of the reaction, but they can be also used for another purpose. For example: One neutron is required to trigger a further fission. Part of free neutrons (0.5 neutrons/fission) is absorbed in other material, but an excess of neutrons (0.9 neutrons/fission) is able to leave the surface of the_reactor core and can be used as a neutron source.

Fusion Systems

Nuclear fusion is a nuclear reaction in which two or more atomic nuclei (*e.g.* D+T) collide at a very high energy and fuse together. Thy byproduct of DT fusion is a free neutron, therefore also nuclear fusion reaction has the potential to produces large quantities of neutrons.

Spallation Sources

Spallation source is a high-flux neutron source in which protons that have been accelerated to high energies hit a heavy target material, causing the emission of neutrons. The reaction occurs above a certain energy threshold for the incident particle, which is typically 5 – 15 MeV.

Medium Neutron Sources

Bremssstrahlung from Electron Accelerators/Photofission

Energetic electrons when slowed down rapidly in a heavy target emit intense gamma radiation during the deceleration process. This is known as Bremsstrahlung or braking radiation. The interaction of the gamma radiation with the target produces neutrons via the (γ,n) reaction, or the (γ,fission) reaction when a fissile target is used. e-→Pb → γ→ Pb →(γ,n) and (γ,fission). The Bremsstrahlung γ energy exceeds the binding energy of the "last" neutron in the target. A source strength of 10^{13} neutrons/second produced in short (*i.e.* < 5 μs) pulses can be readily realised.

Dense Plasma Focus

The dense plasma focus (DPF) is a device that is known as an efficient source of neutrons from fusion reactions. Mechanism of dense plasma focus (DPF) is based on nuclear fusion of short-lived plasma of deuterium and/or tritium. This device produces a short-lived plasma by electromagnetic compression and acceleration that is called **a pinch**. This plasma is during the pinch hot and dense enough to cause nuclear fusion and the emission of neutrons.

Light Ion Accelerators

Neutrons can be also produced by particle accelerators using targets of deuterium, tritium, lithium, beryllium, and other low-Z materials. In this case the target must be bombarded with accelerated hydrogen (H), deuterium (D), or tritium (T) nuclei.

Small Neutron Sources

Neutron Generators

Neutrons are produced in the fusion of deuterium and tritium in the following exothermic reaction.

$$^{2}\mathrm{D} + {}^{3}\mathrm{T} \rightarrow {}^{4}\mathrm{He} + \mathrm{n} + 17.6\ \mathrm{MeV}$$

The neutron is produced with a kinetic energy of 14.1 MeV. This can be achieved on a small scale in the laboratory with a modest 100 kV accelerator for deuterium atoms bombarding a tritium target. Continuous neutron sources of ~10^{11} neutrons/second can be achieved relatively simply.

Radioisotope Source – (α,n) Reactions

In a certain light, isotopes, the 'last' neutron in the nucleus is weakly bound and is released when the compound nucleus is formed following α-particle bombardment decays. The bombardment of beryllium by α-particles leads to the production of neutrons by the following exothermic reaction: $^{4}\mathbf{He} + {}^{9}\mathbf{Be} \rightarrow {}^{12}\mathbf{C} + \mathbf{n} + \mathbf{5.7\ MeV}$. This reaction yields a weak source of neutrons with an energy spectrum resembling that from a fission source and is used nowadays in portable neutron sources. Radium, plutonium or americium can be used as an α-emitter.

Radioisotope Source – (γ,n) Reactions

(γ,n) reactions can also be used for the same purpose. In this type of source, because of the greater range of the γ-ray, the two physical components of the source can be separated, making it possible to 'switch off' the reaction if so required by removing the radioactive source from the beryllium. (γ,n) sources produce a mono-energetic neutrons unlike (α,n) sources. The (γ,n) source uses antimony-124 as the gamma emitter in the following endothermic reaction.

$$^{124}\mathbf{Sb} \rightarrow {}^{124}\mathbf{Te} + \beta- + \gamma$$

$$\gamma + {}^{9}\mathbf{Be} \rightarrow {}^{8}\mathbf{Be} + \mathbf{n} - \mathbf{1.66\ MeV}$$

Radioisotope Source – Spontaneous Fission

Certain isotopes undergo spontaneous fission with emission of neutrons. The most commonly used spontaneous fission source is the radioactive isotope **californium-252**. Cf-252 and all other spontaneous fission neutron sources are produced by irradiating uranium or another transuranic element in a nuclear reactor, where neutrons are absorbed in the starting material and its subsequent reaction products, transmuting the starting material into the SF isotope.

Intrinsic Neutron Sources

Some neutrons will be produced in the materials present in the reactor due to a variety of unavoidable reactions that occur because of the nature of these materials. Intrinsic neutron sources are those neutron-producing reactions that always occur in reactor materials.

A limited number of neutrons will always be present, even in a reactor core that has never been operated, due to spontaneous fission of some heavy nuclides that are present in the fuel. Uranium238, uranium-235, and plutonium-239 undergo spontaneous fission to a limited extent. Uranium238, for example, yields almost 60 neutrons per hour per gram. Table 2.1 illustrates a comparison of the rate at which different heavy nuclides produce neutrons by spontaneous fission.

Table 2.1: Different Rate with which Heavy Nuclides Produce Neutrons

Neutron Production by Spontaneous Fission			
Nuclide	$T_{1/2}$ (Fission)	$T_{1/2}$ (α-decay)	neutrons/sec/gram
$^{235}_{92}U$	1.8×10^{17} years	6.8×10^{8} years	8.0×10^{-4}
$^{238}_{92}U$	8.0×10^{15} years	4.5×10^{9} years	1.6×10^{-2}
$^{239}_{94}Pu$	5.5×10^{5} years	2.4×10^{4} years	3.0×10^{-2}
$^{240}_{94}Pu$	1.2×10^{11} years	6.6×10^{3} years	1.0×10^{3}
$^{252}_{98}Cf$	66.0 years	2.65 years	2.3×10^{12}

Another intrinsic neutron source is a reaction involving natural boron and fuel. In some reactors, natural boron is loaded into the reactor core as a neutron absorber to improve reactor control or increase core life-time. Boron-11 (80.1% of natural boron) undergoes a reaction with the alpha particle emitted by the radioactive decay of heavy nuclides in the fuel to yield a neutron as shown below.

$$^{11}_{5}B + ^{4}_{2}\alpha \rightarrow ^{14}_{7}N + ^{1}_{0}n$$

The boron-11 must be mixed with, or in very close proximity to, the fuel for this reaction because of the short path length of the alpha particle. For a reactor core with this configuration, this (α,n) reaction is an important source of neutrons for reactor startup. In a reactor that has been operated, another source of neutrons becomes significant. Neutrons may be produced by the interaction of a gamma-ray and a deuterium nucleus. This reaction is commonly referred to as a photoneutron reaction because it is initiated by electromagnetic radiation and results in the production of a neutron. The photoneutron reaction is shown below.

$$^{2}_{1}H + ^{0}_{0}\gamma \rightarrow \left(^{2}_{1}H\right)^{*} \rightarrow ^{1}_{1}H + ^{1}_{0}n$$

There is an abundant supply of high energy gammas in a reactor that has been operated because many of the fission products are gamma emitters. All water-cooled reactors have some deuterium present in the coolant in the reactor core because a small fraction of natural hydrogen is the isotope deuterium. The atom percentage of deuterium in the water ranges from close to the naturally occurring value (0.015%) for light water reactors to above 90% deuterium for heavy water reactors. Therefore, the required conditions for production of photoneutrons exist. The supply of gamma-rays decreases with time after shutdown as the gamma emitters decay; therefore, the photoneutron production rate also decreases. In a few particular reactors, additional D_{20} may be added to the reactor to increase the production of photoneutrons following a long shutdown period.

Installed Neutron Sources

Because intrinsic neutron sources can be relatively weak or dependent upon the recent power history of the reactor, many reactors have artificial sources of neutrons installed. These neutron sources ensure that shutdown neutron levels are high enough to be detected by the nuclear instruments at all times. This provides a true picture of reactor conditions and any change in these conditions. An installed neutron source is an assembly placed in or near the reactor for the sole purpose of producing source neutrons.

One strong source of neutrons is the artificial nuclide californium-252, which emits neutrons at the rate of about 2 x 10^{12} neutrons per second per gram as the result of spontaneous fission. Important drawbacks for some applications may be its high cost and its short half-life. Many installed neutron sources use the (α,n) reaction with beryllium. These sources are composed of a mixture of metallic beryllium (100% beryllium-9) with a small quantity of an alpha particle emitter, such as a compound of radium, polonium, or plutonium. The reaction that occurs is shown below.

$$^{9}_{4}\mathrm{Be} + ^{4}_{2}\alpha \rightarrow \left(^{13}_{6}\mathrm{C}\right)^{*} \rightarrow ^{12}_{6}\mathrm{C} + ^{1}_{0}\mathrm{n}$$

The beryllium is intimately (homogeneously) mixed with the alpha emitter and is usually enclosed in a stainless steel capsule. Another type of installed neutron source that is widely used is a photoneutron source that employs the (,n) reaction with beryllium. Beryllium is used for photoneutron sources because its stable isotope beryllium-9 has a weakly attached last neutron with a binding energy of only 1.66 MeV.

Thus, a gamma-ray with greater energy than 1.66 startup sources of this type uses antimony and beryllium as after activation with neutrons the radioactive antimony becomes an emitter of high energy gammas. The photoneutron sources of this type are constructed somewhat differently from the (a,n) types. One design incorporates a capsule of irradiated antimony enclosed in a beryllium sleeve. The entire assembly is then encased in a stainless steel cladding. A large reactor may have several neutron sources of this type installed within the core.

Glossary

- **Background radiation:** Background radiation includes radiation from cosmic sources, naturally occurring radioactive materials (including radon), and global fallout (from the testing of nuclear explosive devices).
- **Cosmic radiation:** Cosmic radiation is penetrating ionizing radiation comprised of particles and electromagnetic energy that comes from outer space.
- **Beta particle:** A beta particle is an energetic electron given off by atoms when the atoms have too much energy.
- **Detector:** A detector is a device or instrument designed to measure whether radioactive material is present.

- **Effective dose:** Radiation exposures to the human body, whether from external or internal sources, can involve all or a portion of the body.
- **Electromagnetic spectrum:** The electromagnetic spectrum is a description of the broad range of radiation that includes radio-waves; microwaves; infrared, visible, and ultraviolet light; and x and gamma-rays.
- **Gamma-rays:** Gamma-rays are high-energy electromagnetic radiation (photons) emitted in an attempt by the radionuclide to become stable, *i.e.*, radioactive decay.
- **Half-life:** Also called radiological or physical half-life, this is the amount of time it takes for half of the radioactivity in a material to be gone or to decay.
- **Ionization:** Ionization is the process by which a neutral atom (an atom with no charge) gains a positive or negative charge.
- **Neutron:** A neutron is an elementary particle that is electrically neutral and found in the nucleus of every atom except hydrogen-1.
- **Personal radiation dosimeter:** More widely known as a radiation "badge," a personal radiation dosimeter is a device worn by an individual to determine the radiation dose received by the individual.
- **Radiation:** Radiation is a term commonly used to describe ionizing radiation (*i.e.*, X and gamma-rays, alpha and beta particles, neutrons).
- **Radioactive decay:** Radioactive decay is the spontaneous transformation of one nuclide into another nuclide or different energy state; in the transformation process, radiation is emitted.
- **Radioactivity:** Radioactivity is the property of a nucleus in unstable atoms that causes them to spontaneously release energy in the form of photons or subatomic particles.
- **Radionuclide:** A radionuclide is a radioactive element, man-made or from natural sources, with a specific atomic weight.
- **Threshold:** Threshold is the point at which radiation first produces an observable effect (response) from acute exposures.
- **X-rays:** X-rays are electromagnetic radiation (photons) that can be emitted from radio-nuclides or from certain types of devices.

3

Basic of Particle Radiation

3.1 Introduction

Radiation is energy that moves through space from one object, *i.e.* the source, to another object where it is absorbed. Radiation sources are generally collections of matter or devices that convert other forms of energy into radiation. In some cases the energy to be converted is stored within the object. Examples are, the sun and radioactive materials. In other cases the radiation source is only an energy converter, and other forms of energy must be applied in order to produce radiation; light bulbs and X-ray tubes are examples.

Particle radiation is an important type of radiation that results from subatomic particles hurtling at tremendous speeds. Particle radiation is around us all the time but it is harder to notice them as compared to the wave radiation. The most common type of particle radiation comes from radioactive atoms such as protons, cosmic rays, alpha particle and beta particles *etc.*

These radiation consists of small particles of matter moving through space at a very high velocity and carry energy as of their motion. Particle radiation comes primarily from radioactive materials, outer space, or machines that accelerate particles to very high velocities, such as linear accelerators, betatrons, and cyclotrons. Particle radiation differs from electromagnetic radiation; in that the particles consist of matter and have mass. The type of particle radiation encountered most frequently in clinical medicine, is high-velocity electron radiation. Particle radiation is generally not used as an imaging radiation because of its low tissue penetration. Also, when X-radiation interacts with matter, such as human tissue, it transfers energy to electrons, thus creating a form of electron radiation within the material. Several types of particle radiation are produced as byproducts of photon production by a number of radioactive materials used in medical imaging.

3.2 Types of Particle Radiation

There are many types of particle radiations such as:

- Protons,
- Cosmic rays
- Alpha particles
- Beta particles

3.2.1 Radioactive Elements

Many nuclei are radioactive; that is, they decompose by emitting particles and in doing so, become a different nucleus. In our studies up to this point, atoms of one element were unable to change into different elements. That is because in all other types of changes we have talked about only the electrons were changing. In these changes, the nucleus, which contains the protons which dictate which element an atom is, is changing. All nuclei with 84 or more protons are radioactive and elements with less than 84 protons have both stable and unstable isotopes. All of these elements can go through nuclear changes and turn into different elements. In natural radioactive decay, three common emissions occur. When these emissions were originally observed, scientists were unable to identify them as some already known particles and so named them-

- Alpha particles (α)
- Beta particles (β)
- Gamma-rays (γ)

Alpha particles were identified as helium-4 nuclei, beta particles were identified as electrons, and gamma-rays as a form of electromagnetic radiation like X-rays except much higher in energy and even more dangerous to living systems.

With all the radiation from natural and man-made sources, we should quite reasonably be concerned about how all the radiation might affect our health. The damage to living systems is done by radioactive emissions when the particles or rays strike tissue, cells, or molecules and alter them. These interactions can alter molecular structure and function; cells no longer carry out their proper function and molecules, such as DNA, no longer carry the appropriate information. Large amounts of radiation are very dangerous, even deadly. In most cases, radiation will damage a single (or very small number) of cells by breaking the cell wall or otherwise preventing a cell from reproducing.

The ability of radiation to damage molecules is analyzed in terms of what is called ionizing power. When a radiation particle interacts with atoms, the interaction can cause the atom to lose electrons and thus become ionized. The greater the likelihood that damage will occur by an interaction is the ionizing power of the radiation.

Much of the threat from radiation is involved with the ease or difficulty of protecting oneself from the particles. How think of wall do you need to hide behind to be safe? The ability of each type of radiation to pass through matter is expressed in terms of penetration power. The more material the radiation can pass through, the greater the penetration power and the more dangerous they are. In general, the greater mass present the greater the ionizing power and the lower the penetration power.

Comparing only the three common types of ionizing radiation, alpha particles have the greatest mass. Alpha particles have approximately four times the mass of a proton or neutron and approximately 8,000 times the mass of a beta particle. Because of the large mass of the alpha particle, it has the highest ionizing power and the greatest ability to damage tissue. That same large size of alpha particles, however, makes them less able to penetrate matter. They collide with molecules very quickly when striking matter, add two electrons, and become a harmless helium atom.

Alpha particles have the least penetration power and can be stopped by a thick sheet of paper or even a layer of clothes. They are also stopped by the outer layer of dead skin on people. This may seem to remove the threat from alpha particles but only from external sources. In a situation like a nuclear explosion or some sort of nuclear accident where radioactive emitters are spread around in the environment, the emitters can be inhaled or taken in with food or water and once the alpha emitter is inside you, you have no protection at all.

Beta particles are much smaller than alpha particles and therefore, have much less ionizing power, but their small size gives them much greater penetration power. Most resources say that beta particles can be stopped by a one-quarter inch thick sheet of aluminum.

Gamma-rays are not particles but a high energy form of electromagnetic radiation (like X-rays except more powerful). Gamma-rays are energy that has no mass or charge. Gamma-rays have tremendous penetration power and require several inches of dense material to shield them. Gamma-rays may pass all the way through a human body without striking anything. They are considered to have the least ionizing power and the greatest penetration power.

Decay

Some isotopes of certain elements have an atomic structure that is naturally unstable. This is where the nucleus of an atom is unstable because it has the wrong number of neutrons relative to its number of protons. Radioactive decay is the way that these unstable atoms emit or radiate excess energy in the form of particles or waves. Depending on how the nucleus loses this excess energy, it may form a lower energy isotope of the same element. Or it could lose enough particles to become a different element altogether.

Decay of a radioactive nucleus results in stability. Often, a radioactive nucleus cannot reach a stable state through a single decay. In such cases, a series of decays will occur until a stable nucleus is formed. The decay of UU-238 is an example of this (Figure 3.1). The UU-238 decay series starts with UU-238 and goes through fourteen separate decays to finally reach a stable nucleus, PbPb-206. There are similar decay series for UU-235 and ThTh-232. The UU-235 series ends with PbPb-207 and the ThTh-232 series ends with PbPb-208.

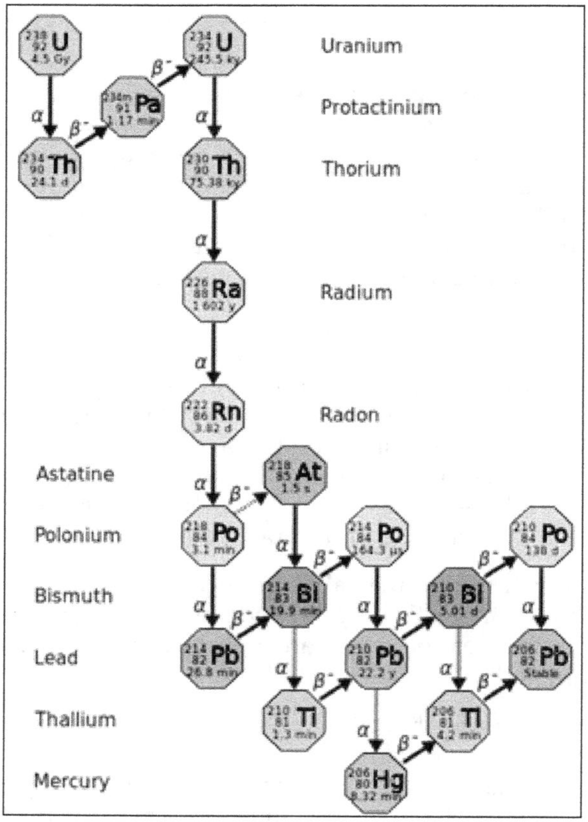

Figure 3.1: Uranium-238 Decay Chain.

Several of the radioactive nuclei that are found in nature are present there because they are produced in one of the radioactive decay series. For example, there may have been radon on the earth at the time of its formation, but that original radon would have all decayed by this time. The radon that is present now is present because it was formed in a decay series.

Alpha Decay

Alpha decay is one process that unstable atoms can use to become more stable. During alpha decay, an atom's nucleus sheds two protons and two neutrons in a packet that scientists call an alpha particle. Since an atom loses two protons during alpha decay, it changes from one element to another. For example, after undergoing alpha decay, an atom of uranium (with 92 protons) becomes an atom of thorium (with 90 protons).

Alpha decay is the spontaneous radioactive decay where an alpha particle is produced. An alpha particle is essentially a helium nucleus or He^{2+} ion. Although alpha decay present a significant radiation risk if the radioactive source is inhaled or ingested, alpha particles are too large to penetrate very far through skin or other solids and require minimal radiation shielding. The spontaneous radioactive decay is depicted in Figure 3.2.

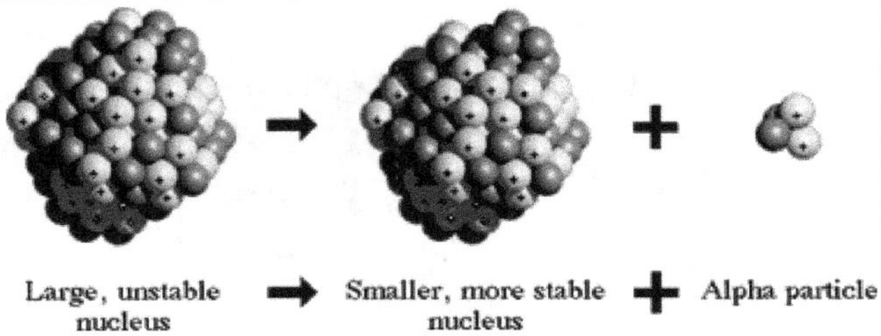

Figure 3.2: Spontaneous Radioactive Decay.

An atom that undergoes alpha decay will reduce its atomic mass by 4 and become the element two atomic numbers less. The general reaction for alpha decay is:

$$^{Z}X_{A} \rightarrow {}^{Z-4}Y_{A-2} + {}^{4}He_{2}$$

where

X = parent atom
Y = daughter atom
Z = atomic mass of X
A = atomic number of X

Many types of heavy atomic nucleus spontaneously decay to produce daughter nuclei via the emission of α-particles (*i.e.*, helium nuclei) of some characteristic energy. This process is known as α-decay. Let us investigate the α-decay of a particular type of atomic nucleus of radius R, charge-number Z, and mass-number *A*. Such a nucleus, thus decays to produce a daughter nucleus of charge-number Z1=Z-2 and mass-number A1=A-4, and an α-particle of charge-number Z_2=2 and mass-number A_2=4. Let the characteristic energy of the α-particle be *E*. Incidentally, nuclear radii are found to satisfy the empirical formula:

$$R = 1.5 \times 10–15\ A1/3\ m = 2.0 \times 10–15\ Z_1^{1/3} m$$

for Z>>1

In 1928, George Gamov proposed a very successful theory of α-decay, according to which the α-particle moves freely inside the nucleus, and is emitted after *tunneling* through the potential barrier between itself and the daughter nucleus. In other words, the α-particle, whose energy is *E*, is trapped in a potential well of radius *R* by the potential barrier

$$V(r) = \frac{Z_1 Z_2 e^2}{4\pi \epsilon_0 r}$$

for r>R

Making use of the WKB approximation (and neglecting the fact that γ is a radial, rather than a Cartesian, coordinate), the probability of the α-particle tunneling through the barrier is

$$|T|^2 = \exp\left(-\frac{2\sqrt{2m}}{\hbar}\int_{r1}^{r2}\sqrt{V(r)-E\,dr}\right),$$

Where, r_1=R and r_2=$Z_1Z_2e^2/(4\pi\varepsilon_o E)$. Here, m=4mp is the α-particle mass. The above expression reduces to

$$|T|^2 = \exp\left(-2\sqrt{2}\beta\int_1^{E_c/E}\left[\frac{1}{y}-\frac{E}{E_c}\right]^{1/2}dy\right),$$

Where

$$\beta = \left(\frac{Z_1 Z_2 e^2 m R}{4\pi\epsilon_0\hbar^2}\right)^{1/2} = 0.74\,Z_1^{2/3}$$

is a dimensionless constant, and

$$E_c = \frac{Z_1 Z_2 e^2}{4\pi\epsilon_0 R} = 1.44\,Z_1^{2/3}\text{ MeV}$$

is the characteristic energy the α-particle would need in order to escape from the nucleus without tunneling, E<<E_c. It is easily demonstrated that

$$\int_1^{1/\epsilon}\left[\frac{1}{y}-\epsilon\right]^{1/2}dy \simeq \frac{\pi}{2\sqrt{\epsilon}}-2$$

when ∈» 1.

Hence,

$$|T|^2 \simeq \exp\left(-2\sqrt{2}\beta\left[\frac{\pi}{2}\sqrt{\frac{E_c}{E}}-2\right]\right).$$

Now, the α-particle moves inside the nucleus with the characteristic velocity v = √2E/m. It follows that the particle bounces backward and forward within the nucleus at the frequency v ~ v/R, giving v ~ 2×10^{28} yr for a 1 MeV α-particle trapped inside a typical heavy nucleus of radius 10^{-14} m. Thus, the α-particle effectively attempts to tunnel through the potential barrier ν times a second. If each of these attempts has a probability $|T|^2$ of succeeding, then the probability of decay per unit time is $v|T|^2$. Hence, if there are N(t)>>1 undecayed nuclei at time t then there are only N+dN at time t+dt, where

$$dN = -N\nu\,|T|^2\,dt.$$

This expression can be integrated to give

$$N(t) = N(0)\, \exp(-\nu\, |T|^2\, t).$$

Now, the *half-life*, τ, is defined as the time which must elapse in order for half of the nuclei originally present to decay. It follows from the above formula that

$$\tau = \frac{\ln 2}{\nu\, |T|^2}.$$

Note that the half-life is *independent* of N(0). Finally, making use of the above results, we obtain

$$\log_{10}[\tau(\mathrm{yr})] = -C_1 - C_2\, Z_1^{2/3} + C_3\, \frac{Z_1}{\sqrt{E(\mathrm{MeV})}},$$

Where

C1 = 28.5
C2 = 1.83
C3 = 1.73

$$\log_{10}(\tau_{th}) = -28.9 - 1.60\, Z_1^{2/3} + 1.61\, Z_1/\sqrt{E}$$

The half-life, τ, the daughter charge-number, Z1=Z-2, and the α-particle energy, E, for atomic nuclei which undergo α-decay are indeed found to satisfy a relation (Figure 3.3).

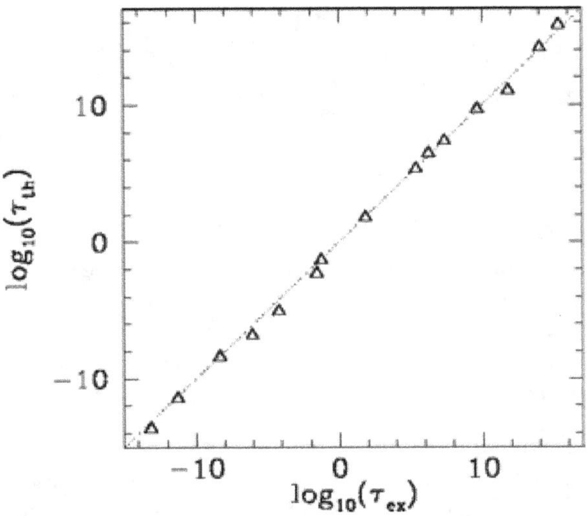

Figure 3.3: Experimentally Determination of Half-life, T_{ex}, of Atomic Nucleii that Decay by α-emission.

The best fit to the data is obtained using

C1 = 28.9
C2 = 1.60
C3 = 1.61

Beta Decay

Beta particles are electrons or positrons (electrons with positive electric charge, or anti-electrons). Beta decay occurs when, in a nucleus with too many protons or too many neutrons, one of the protons or neutrons is transformed into the other. In beta minus decay, a neutron decays into a proton, an electron, and an anti-neutrino: n Æ p + e - +. In beta plus decay, a proton decays into a neutron, a positron, and a neutrino: p Æ n + e+ +n. Both reactions occur because in different regions of the Chart of the Nuclides, one or the other will move the product closer to the region of stability. These particular reactions take place because conservation laws are obeyed.

Electric charge conservation requires that if an electrically neutral neutron becomes a positively charged proton, an electrically negative particle (in this case, an electron) must also be produced. Similarly, conservation of lepton number requires that if a neutron (lepton number = 0) decays into a proton (lepton number = 0) and an electron (lepton number = 1), a particle with a lepton number of -1 (in this case an anti-neutrino) must also be produced. The leptons emitted in beta decay did not exist in the nucleus before the decay–they are created at the instant of the decay (Figure 3.4).

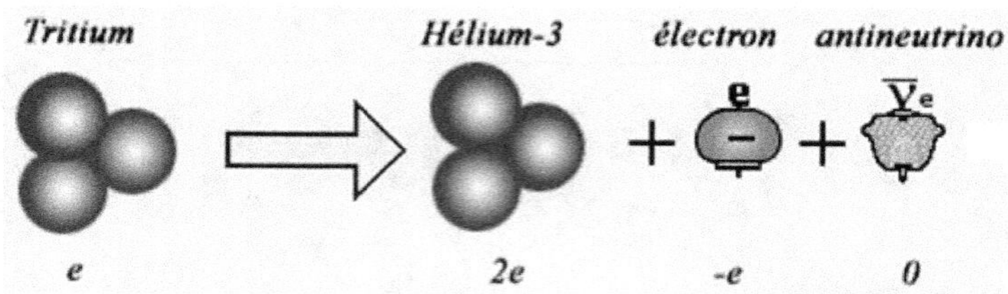

Figure 3.4: Beta Decay.

Proton decay, neutron decay, and electron capture are three ways in which protons can be changed into neutrons or vice-versa; in each decay there is a change in the atomic number, so that the parent and daughter atoms are different elements. In all three processes, the number A of nucleons remains the same, while both proton number, Z, and neutron number, N, increase or decrease by 1.

In beta decay the change in binding energy appears as the mass energy and kinetic energy of the beta particle, the energy of the neutrino, and the kinetic energy of the recoiling daughter nucleus. The energy of an emitted beta particle from a particular decay can take on a range of values because the energy can be shared in many ways among the three particles while still obeying energy and momentum conservation.

When a beta particle is emitted from the nucleus; the nucleus has one more proton and one less neutron. This means the atomic mass number remains unchanged and the atomic number increases by 1.

Example:

Carbon-14 is a radioactive isotope of carbon having 8 neutrons instead of 6. Here is the equation for the beta decay of carbon-14 into nitrogen.

$$^{14}_{6}C \rightarrow ^{14}_{7}N + ^{0}_{-1}e^{-}$$

Unlike β- decay, which occurs when radioisotopes have too many neutrons, β+ decay occurs when they have too many protons. In this case, a proton is converted into a neutron and a positive beta particle of β+. This is called a positron and has the same mass as an electron but the opposite charge (positive).

β^- Decay (Electron Emission)

An unstable atomic nucleus with an excess of neutrons may undergo β– decay. In this process, a neutron is converted into a proton, an electron, and an electron-type anti-neutrino (the antiparticle of the neutrino):

$$n^0 \rightarrow p^+ + e^- + \bar{\nu}.$$

At the fundamental level, this process is mediated by the weak interaction. A neutron (one up quark and two down quarks) turns into a proton (two up quarks and one down quark) by the conversion of a down quark to an up quark, with the emission of a W^- boson. The W^- boson subsequently decays into an electron and an anti-neutrino.

Beta decay commonly occurs among the neutron-rich fission byproducts produced in nuclear reactors. This process is the source of the large numbers of electron anti-neutrinos produced by fission reactors. Free neutrons also decay via this process.

β^+ Decay (Positron Emission)

Unstable atomic nuclei with an excess of protons may undergo β+ decay, or inverse beta decay. In this case, energy is used to convert a proton into a neutron, a positron (e+), and an electron-type neutrino (ν_e):

$$\text{energy} + p+ \rightarrow n^0 + ve^+ + v_e.$$

On a fundamental level, an up quark is converted into a down quark, emitting a W+ boson that then decays into a positron and a neutrino.

Unlike beta minus decay, beta plus decay cannot occur in isolation, because it requires energy — the mass of the neutron being greater than the mass of the proton. Beta plus decay can only happen inside nuclei when the absolute value of the binding energy of the daughter nucleus is higher than that of the mother nucleus. The difference between these energies goes into the reaction of converting a proton into a neutron, a positron and, a neutrino and into the kinetic energy of these particles.

Beta decay does not change the number of nucleons A in the nucleus, but changes only its charge Z. Thus, during beta decay, the parent nuclide and daughter nuclide share the same A value. The beta decay of atomic nuclei results in the transmutation of one chemical element into another. For example:

Beta minus: $\frac{137}{55}\text{Cs} \rightarrow \frac{137}{56}\text{Ba} + e^- + \bar{v}$

Beta plus: $\frac{22}{11}\text{Na} \rightarrow \frac{22}{10}\text{Ne} + e^+ + v_e$

For comparison, the electron capture process may be written as follows:

Electron capture: $\frac{22}{11}\text{Na} + e^+ \rightarrow \frac{22}{10}\text{Ne} + v_e$

In nature, most isotopes are beta stable, but a few exceptions exist with half-lives so long that they have not had enough time to decay since the moment of their nucleosynthesis. One example is ^{40}K, which undergoes beta minus and beta plus decay and electron capture, with a half-life of 1.277×10^9 years.

It should be noted that a beta-stable nucleus may undergo other kinds of radioactive decay, such as alpha decay.

Gamma Radiation

Gamma decay is one type of radioactive decay that a nucleus can undergo. What separates this type of decay process from alpha or beta decay is that no particles are ejected from the nucleus when it undergoes this type of decay. Instead, a high energy form of electromagnetic radiation - a gamma-ray photon - is released. Gamma-rays are simply photons that have extremely high energies which are highly ionizing.

In gamma decay, as shown in Figure 3.5, a nucleus changes from a higher energy state to a lower energy state through the emission of electromagnetic radiation (photons). The number of protons (and neutrons) in the nucleus does not change in this process, so the parent and daughter atoms are the same chemical element. In the gamma decay of a nucleus, the emitted photon and recoiling nucleus each have a well-defined energy after the decay. The characteristic energy is divided between only two particles.

Figure 3.5: Gamma Decay.

Gamma radiation is unique in the sense that undergoing gamma decay does not change the structure or composition of the atom. Instead, it only changes the energy of the atom since the gamma-ray carries no charge nor does it have an associated mass (Figure 3.6).

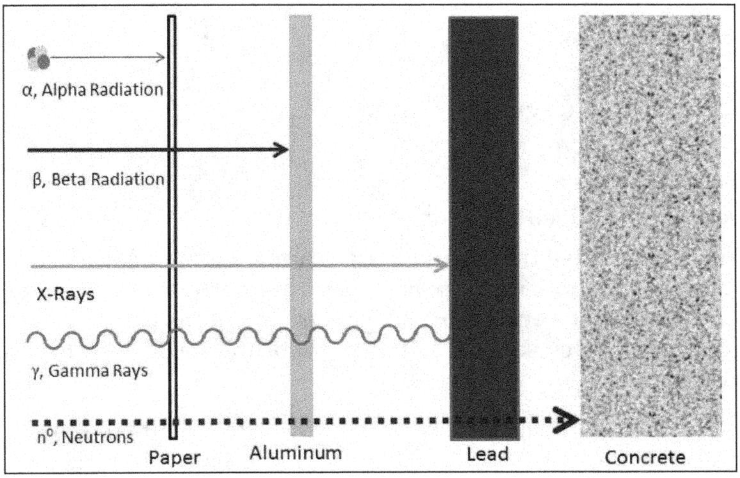

Figure 3.6: Different Penetration Levels of Products of Decay.

In order for a nucleus to undergo gamma decay, it must be in some sort of excited energetic state. Experiments have shown that protons and neutrons are located in discrete energy states within the nucleus, not too different from the excited states that electrons can occupy in atoms. Thus if a proton or a neutron inside of the nucleus jumps up to an excited state - generally following an alpha or beta decay - the new daughter nucleus must somehow release energy to allow the proton or neutron to relax back down to ground state (Figure 3.7). When the nucleon makes this transition from a high to a low energy state, a gamma photon is emitted. The general equation that represents this process is:

$$A^* \rightarrow A + \gamma$$

where:

- A^* is the excited atom
- A is a relaxed state of the initial excited atom
- γ is the released gamma-ray photon

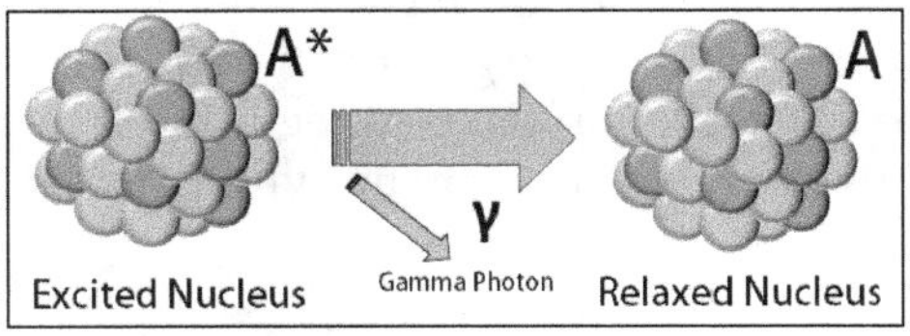

Figure 3.7: Gamma Decay of Nucleus.

Knowing that an atom undergoes gamma radiation is important, but it is also possible to determine the frequency of the released gamma radiation if the initial and final states of the nucleon inside the nucleus are known. The equation representing the frequency of the gamma radiation is:

$E_i - E_f = hf$

where:

- E_i = initial, higher energy state of nucleon
- E_f = final, lower energy state of nucleon
- h = Planck's constant
- f = frequency of emitted radiation

In addition to radioactive nuclei, there are many objects in space that emit high levels of gamma radiation. Gamma-rays can at times be harmful due to the fact that they are generally very high energy and therefore penetrate matter very easily. Since it penetrates so easily, it is some of the most useful radiations for medical purposes.

Some of the most widely used gamma emitters are cobalt-60, cesium-137, and technetium-99m. Cesium is used widely in radiotherapy - the treatment of cancer using gamma-rays - as well as being used to measure soil density at construction sites and to investigate the subterranean layers of the Earth in oil wells. Cobalt is used to sterilize medical equipment and irradiate food, killing bacteria and pasteurizing the food. Technetium-99m (which has a shorter half-life than technetium-99) is the most widely used for diagnostic medical tests to investigate the brain, bone, and internal organs. As well, exposure to gamma radiation can improve the durability of wood and plastics, and is thus used to toughen flooring in high-traffic areas.

In addition, uranium-238 and uranium-235 - used in fuel for nuclear power plants - undergo both alpha and gamma decays when used. Immediately following the fission process, gamma-rays are released, resulting in high levels of radiation present around the reactor. However, safety precautions are in line to ensure that workers do not get close enough to this radioactive area to be harmed.

3.2.2 Cosmic Rays

Cosmic rays are a type of radiation that comes from space. Cosmic rays aren't really "rays". They are particles (mostly protons) with very high energies. Cosmic rays come from various places, including the Sun, supernova explosions, and extremely distant sources such as radio galaxies and quasars. Because of their high energy, this type of particle radiation can be dangerous to people and to machines. On Earth we are mostly shielded from them by our planet's magnetic field and atmosphere.

Cosmic ray is the term given to high energy radiation which strikes the Earth from space. Some of them have ultrahigh energies in the range 100 - 1000 TeV. Such extreme energies come from only a few sources like Cygnus X-3. The peak of the energy distribution is at about 0.3 GeV. The intensity of cosmic radiation increases with altitude, indicating that it comes from outer space. It changes with latitude, indicating that it consists at least partly of charged particles which are affected by the earth's magnetic field. The illustration at right shows that the detected cosmic ray flux peaks at about 15 km in altitude and then drops sharply (Figure 3.8). This kind of variation was discovered by Pfotzer in 1936. It suggests that the detection method used was mainly detecting secondary particles rather than the primary particles reaching the Earth from space.

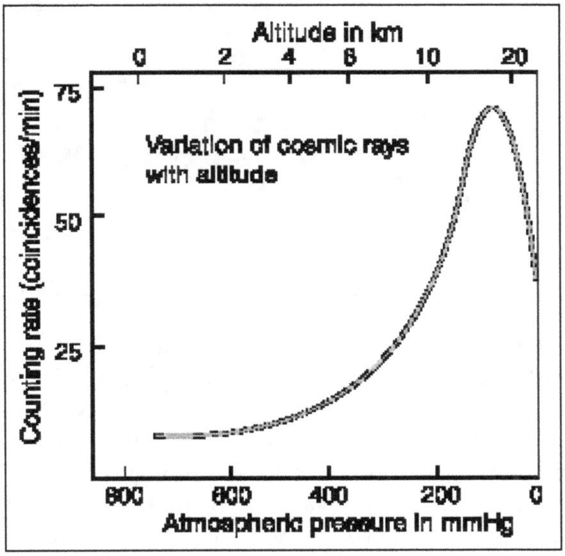

Figure 3.8: Logarithmic Scale on the Altitude.

Cosmic rays were discovered in 1912 by Victor Hess, an Austrian physicist who took an ionization chamber (a device which detects charged particles) up into the Earth's atmosphere in a balloon. As the balloon rose he found that the numbers of charged particles initially dropped off, easily explained if these particles came from the Earth. Then, surprisingly, the numbers of charged particles began to rise again. He concluded that these charged particles were coming from outside the Earth's atmosphere and named them cosmic rays. In 1936 he was awarded the Nobel Prize in physics for this work.

They were initially thought to be some sort of electromagnetic radiation, *i.e.*, made up of photons, as X-rays or gamma-rays are. With the first measurements, made in 1912 by Victor Hess with a balloon, we learned that "a radiation of very great penetrating power enters our atmosphere from above," which excluded the Sun as its source. A few years later, we discovered that this "radiation" was in fact charged particles, since we observed that cosmic rays are deflected by magnetic fields. Photons are neutral particles and are not deflected by magnetic fields (Figure 3.9).

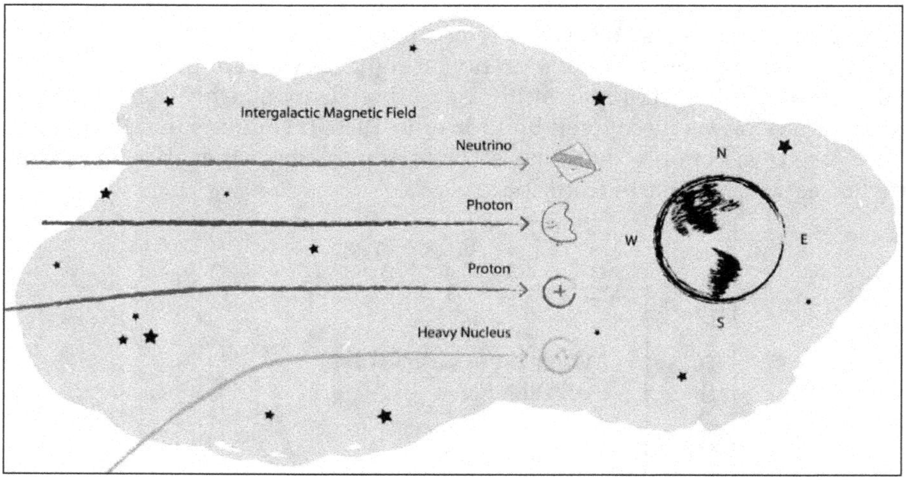

Figure 3.9: Bending of Charged Particles when Travel through Magnetic Fields.

There are several different types of cosmic rays. Solar cosmic rays come from the Sun. They have less energy than most other types of cosmic rays. Solar flares and other explosions on the Sun make this type of cosmic ray. A second type is the galactic cosmic ray. Galactic cosmic rays have more energy than solar cosmic rays. Astronomers think they come from supernova explosions, black holes, and neutron stars within our own Milky Way galaxy. A third type is the rare extragalactic cosmic ray. Scientist think these particles come from somewhere outside of our galaxy. However, they aren't exactly sure where they come from. Extragalactic cosmic rays have even more energy than galactic cosmic rays. Extragalactic cosmic rays may come from quasars or from the nucleus of an active galaxy. Maybe they are made when galaxies crash into each other. Nobody knows for sure. Finally, there are the strange anomalous cosmic rays (ACR). ACRs have lower energies than other types. Astronomers think they might come from the edge of our Solar System. The Sun's magnetic field makes a gigantic "bubble" in space called the heliosphere. ACRs may come from the edge of the heliosphere, where the Sun's magnetic field bumps into interstellar space.

Since cosmic rays are a kind of radiation, they can hurt people and machines. Lucky for us, Earth's magnetic field and atmosphere protect us from most cosmic rays. On an average, about 2.3 millisieverts of radiation strike people each year. A millisievert is a unit for measuring radiation. It is abbreviated as mSv. Cosmic rays make up about 0.2 mSv of the radiation we get each year. That isn't very much; less than 10% of the total. Astronauts do have to worry about cosmic rays, though. If astronauts travel away from Earth (to the Moon or Mars, for example), they aren't protected by Earth's magnetic field any more. They could get hit by as much as 900 mSv of cosmic ray radiation in a year! Cosmic rays can damage our DNA and cause cancer and radiation sickness. Scientists will have to figure out how to protect astronauts from cosmic rays before we can send a mission to Mars.

When cosmic rays hit Earth's atmosphere, they crash into atoms and molecules of gas. That usually makes even more cosmic ray particles! Since there are more particles, the energy from the cosmic ray from space is spread out. The new cosmic ray particles often hit other gas molecules. That makes still more cosmic rays, but with lower energies. The collisions between cosmic rays and gases in the atmosphere can happen many times. In the end, there might be thousands or millions of "secondary" cosmic rays. This is called an "air shower" of cosmic rays.

Cosmic Ray Energies and Acceleration

The energy of cosmic rays is usually measured in units of MeV, for mega-electron volts, or GeV, for giga-electron volts. (One electron volt is the energy gained when an electron is accelerated through a potential difference of 1 volt). Most galactic cosmic rays have energies between 100 MeV (corresponding to a velocity for protons of 43% of the speed of light) and 10 GeV (corresponding to 99.6% of the speed of light). The number of cosmic rays with energies beyond 1 GeV decreases by about a factor of 50 for every factor of 10 increase in energy. Over a wide energy range the number of particles per m^2 per steradian per second with energy greater than E (measured in GeV) is given approximately by $N(>E) = k(E + 1)^{-a}$, where $k \sim 5000$ per m^2 per steradian per second and $a \sim 1.6$. The highest energy cosmic rays measured to date have had more than 10^{20} eV, equivalent to the kinetic energy of a baseball traveling at approximately 100 mph.

It is believed that most galactic cosmic rays derive their energy from supernova explosions, which occur approximately once every 50 years in our Galaxy. To maintain the observed intensity of cosmic rays over millions of years, requires, that a few percent of the more than 10^{51} ergs released in a typical supernova explosion be converted to cosmic rays. There is considerable evidence that cosmic rays are accelerated as the shock waves from these explosions travel through the surrounding interstellar gas. The energy contributed to the Galaxy by cosmic rays (about 1 eV per cm^3) is about equal to that contained in galactic magnetic fields, and in the thermal energy of the gas that pervades the space between the stars.

Cosmic Ray Composition

Cosmic rays include essentially all of the elements in the periodic table; about 89% of the nuclei are hydrogen (protons), 10% helium, and about 1% heavier elements. The common heavier elements (such as carbon, oxygen, magnesium, silicon, and iron) are present in about the same relative abundances as in the solar system, but there are important differences in elemental and isotopic composition that provide information on the origin and history of galactic cosmic rays. For example there is a significant overabundance of the rare elements Li, Be, and B produced when heavier cosmic rays such as carbon, nitrogen, and oxygen fragment into lighter nuclei during collisions with the interstellar gas. The isotope ^{22}Ne is also overabundant, showing that the nucleo-synthesis of cosmic rays and solar system material have differed. Electrons constitute about 1% of galactic cosmic rays. It is not known why electrons are apparently less efficiently accelerated than nuclei.

Cosmic Rays in the Galaxy

As cosmic rays are electrically charged they are deflected by magnetic fields, and their directions have been randomized, making it impossible to tell where they originated. However, cosmic rays in other regions of the Galaxy can be traced by the electromagnetic radiation they produce. Supernova remnants such as the Crab Nebula are known to be a source of cosmic rays from the radio synchrotron radiation emitted by cosmic ray electrons spiraling in the magnetic fields of the remnant. In addition, observations of high energy (10 MeV - 1000 MeV) gamma-rays resulting from cosmic ray collisions with interstellar gas show that most cosmic rays are confined to the disk of the Galaxy, presumably by its magnetic field. Similar collisions of cosmic ray nuclei produce lighter nuclear fragments, including radioactive isotopes such as ^{10}Be, which has a half-life of 1.6 million years. The measured amount of ^{10}Be in cosmic rays implies that, on average, cosmic rays spend about 10 million years in the Galaxy before escaping into inter-galactic space.

Very High Energy Cosmic Rays

When high energy cosmic rays undergo collisions with atoms of the upper atmosphere, they produce a cascade of "secondary" particles that shower down through the atmosphere to the Earth's surface. Secondary cosmic rays include pions (which quickly decay to produce muons, neutrinos and gamma-rays), as well as electrons and positrons produced by muon decay and gamma-ray interactions with atmospheric atoms. The number of particles reaching the Earth's surface is related to the energy of the cosmic ray that struck the upper atmosphere. Cosmic rays with energies beyond 10^{14} eV are studied with large "air shower" arrays of detectors distributed over many square kilometers that sample the particles produced. The frequency of air showers ranges from about 100 per m^2per year for energies $>10^{15}$ eV to only about 1 per km^2 per century for energies beyond 10^{20} eV. Cosmic ray interaction products such as neutrinos are also studied by large detectors placed deep in underground mines or under water.

Most secondary cosmic rays reaching the Earth's surface are muons, with an average intensity of about 100 per m^2 per second. Although thousands of cosmic rays pass through our bodies every minute, the resulting radiation levels are relatively low, corresponding, and at sea level, to only a few percent of the natural background radiation. However, the greater intensity of cosmic rays in outer space is a potential radiation hazard for astronauts, especially when the Sun is active, and interplanetary space may suddenly be filled with solar energetic particles.

Cosmic rays are also a hazard to electronic instrumentation in space; impacts of heavily-ionizing cosmic ray nuclei can cause computer memory bits to "flip" or small microcircuits to fail.

Cosmic Rays in the Solar System

Just as cosmic rays are deflected by the magnetic fields in interstellar space, they are also affected by the interplanetary magnetic field embedded in the solar wind (the plasma of ions and electrons blowing from the solar corona at about 400 km/sec), and therefore have difficulty reaching the inner solar system. Spacecraft venturing out towards the boundary of the solar system they have found that the intensity of galactic cosmic rays increases with distance from the Sun. As solar activity varies over the 11 year solar cycle; the intensity of cosmic rays at Earth also varies, in anti-correlation with the sunspot number.

The Sun is also a sporadic source of cosmic ray nuclei and electrons that are accelerated by shock waves traveling through the corona, and by magnetic energy released in solar flares. During such occurrences the intensity of energetic particles in space can increase by a factor of $10^2$to 10^6 for hours to days. Such solar particle events are much more frequent during the active phase of the solar cycle. The maximum energy reached in solar particle events is typically 10 to 100 MeV, occasionally reaching 1 GeV (roughly once a year) to 10 GeV (roughly once a decade). Solar energetic particles can be used to measure the elemental and isotopic composition of the Sun, thereby complementing spectroscopic studies of solar material.

Glossary

- **Absorber:** Any material that stops ionizing radiation. Lead, concrete, and steel attenuate gamma-rays. A thin sheet of paper or metal will stop or absorb alpha particles and most beta particles.
- **Accelerator:** Device used to increase the energy of particles, which then collide with other particles. Major types are linear accelerators and circular accelerators. The name refers to the path taken by the accelerated particle.
- **Activity:** The rate of radioactive decay.
- **Alpha particle** (alpha radiation, alpha ray): A 4He nucleus. It is made up of two neutrons and two protons. It is the least penetrating of the three common forms of radiation, being stopped by a thin sheet of paper. It is not dangerous to living things unless the alpha-emitting substance is inhaled or ingested or comes into contact with the lens of the eye.
- **Annihilation:** Annihilation of particles is the disappearance of the mass energy of a particle and its corresponding antiparticle, and its appearance as another sort of energy (possibly including a spray of particles of total quantum number zero for each of the additive quantum numbers).
- **Anti-neutrino:** Antiparticle to the neutrino. See antiparticles.

- **Antiparticle:** Particle having the same mass, spin, isospin as a particle, but having all additive quantum numbers opposite to those of its respective particle. Antiparticles have the opposite charge of its corresponding particle. Antibaryons are antiparticles to baryons, antileptons are antiparticles to leptons, antiquarks are antiparticles to quarks. The antiparticle for a particular particle, for example a neutrino, is denoted an anti-neutrino.
- **Asymptotic freedom:** Quark-quark interactions weaken as the energy gets higher, or, equivalently, as the quarks approach one another.
- **Atom:** A particle of matter indivisible by chemical means. It is the fundamental building block of molecules. It consists of a positively charged nucleus and orbiting electrons. The number of electrons is the same as the number of protons in the nucleus.
- **Atomic mass** (sometimes mistakenly called atomic weight): The mass of a neutral atom. Its value in atomic mass units (u) is approximately equal to the sum of the number of protons and neutrons in the nucleus of the atom.
- **Atomic mass number:** A, the total number of nucleons (protons and neutrons) found in a nucleus.
- **Atomic number:** Z, the total number of protons found in a nucleus.
- **Atomic mass unit (amu or u):** Unit of mass defined by the convention that the atom 12C has a mass of exactly 12 u; the mass of 1 u is 1.67 ′ 10-27 kg.
- **Background radiation:** The radiation found in the natural environment originating primarily from the naturally radioactive elements of Earth and from cosmic rays. The term may also mean radiation extraneous to an experiment.
- **Baryon:** A massive composite hadron (made of three quarks) such as the proton or the neutron.
- **Baryon number:** Quantum number characteristic of baryons. Each baryon has a value of +1, while each anti-baryon has a value of -1.
- **Becquerel (Bq):** Unit of activity in the International System–one disintegration per second; 1 Bq = 27 pci.
- **Beta particle (beta radiation, beta ray):** An electron of either positive charge (e+ or b+) or negative charge (e, e- or b-) emitted by an atomic nucleus or neutron in the process of a transformation. Beta particles are more penetrating than alpha particles but less than gamma-rays or X-rays. Electron capture is a form of beta decay.
- **Big Bang:** Beginning of the universe; a transition from conditions of unimaginable density and temperature to conditions of lower density and temperature.
- **Blackbody:** An object that is a perfect emitter and absorber of radiation.

- **Blackbody radiation:** Radiation emitted by a blackbody (the intensity depends on temperature).
- **Black hole:** An object so dense that light cannot escape from it.
- **Boson:** A particle having spin that is an integer multiple of (h-bar).
- Cherenkov radiation: Light emitted by particles that move through a medium in which the speed of light is slower than the speed of the particles.
- **Conservation law:** A relation asserting that a specific quantity is conserved. For example, conservation of energy, conservation of momentum, conservation of electron number. Conservation laws are connected to symmetries through Noether's theorem.
- **Curie (Ci):** The original unit used to describe the intensity of radioactivity in a sample of material. One curie equals thirty-seven billion disintegrations per second, or approximately the radioactivity of one gram of radium. This unit is no longer recognized as part of the International System of units. It has been replaced by the becquerel.
- **Decay (radioactive):** The change of one radioactive nuclide into a different nuclide by the spontaneous emission of radiation such as alpha, beta, or gamma-rays, or by electron capture. The end product is a less energetic, more stable nucleus. Each decay process has a definite half-life.
- **Decay rate:** The ratio of activity to the number of radioactive atoms of a particular species.
- **Decay time:** The time required for a quantity to fall to 1/e times the original value.
- **Density:** The ratio of an object's mass to its volume.
- **Detector:** A device or series of devices to used to measure nuclear particles and radiations.
- **Dose:** A general term denoting the effect of absorption of a quantity of radiation or energy absorbed.
- **Electric dipole moment:** The product of charge and distance of separation for an electric dipole.
- **Electromagnetic radiation:** Radiation consisting of electric and magnetic fields that travel at the speed of light. Examples: light, radio waves, gamma-rays, X-rays.
- **Electron:** An elementary particle with a unit electrical charge and a mass 1/1837 that of the proton. Electrons surround an atom's positively charged nucleus and determine that atom's chemical properties.

- **Electron capture:** A radioactive decay process in which an orbital electron is captured by and merges with the nucleus. The mass number is unchanged, but the atomic number is decreased by one.
- **Electroweak interaction:** A theory which unifies the electromagnetic and weak interactions.
- **Excited state:** The state of an atom or nucleus when it possesses more than its normal energy. Typically, the excess energy is released as a gamma-ray.
- **Fermion:** A particle having a spin that is an odd integer multiple of (h-bar)/2.
- **Fissile nucleus:** A nucleus that may fission after collision with a thermal (slow) neutron or that fissions spontaneously (by itself).
- **Fission:** The splitting of a heavy nucleus into two roughly equal parts (which are nuclei of lower-mass elements), accompanied by the release of a relatively large amount of energy in the form of kinetic energy of the two parts and in the form of emission of neutrons and gamma-rays.
- **Fission products:** Nuclei formed by the fission of higher mass elements. They are of medium atomic mass and almost all are radioactive. Examples: 90Sr, 137Ce.
- **Fusion:** A process whereby low mass nuclei combine to form a more massive nucleus plus one or move massive particles.
- **Gamma-ray:** A highly penetrating type of nuclear radiation, similar to X-radiation, except that it comes from within the nucleus of an atom, and, in general, has a shorter wavelength.
- **Gauge boson:** Particle mediating an interaction. By exchange of the gauge particle, the interaction between two particles is accomplished.
- **Geiger counter:** A Geiger-Müller detector and measuring instrument. It consists of a gas-filled tube that discharges electrically when ionizing radiation passes through it and a device that records the events.
- **Gluon:** A gauge particle mediating the color strong interaction.
- **Gray (Gy):** Unit of absorbed dose due to any type of radiation. An exposure to 1 gray results from radiation depositing one joule per kilogram of animal tissue or any other material.
- **Hadron:** A strongly interacting particle.
- **Half-life:** The time in which half the (large number of) atoms of a particular radioactive nuclide disintegrate. The half-life is a characteristic property of each radioactive isotope.
- **Homolog (or homologs):** Elements in the same periodic table group that tend to exhibit similar, but not identical, chemical properties.

- **Hormesis:** Controversial theory which argues that there is a benefit to health, or decrease in biological damage from radiation as dose in increased (valid only for very small doses).
- **Hubble Constant:** Ratio of outward speed of galaxies to their distances from Earth.
- **Induced radioactivity:** Radioactivity that is created by bombarding a substance with neutrons in a reactor or with charged particles produced by particle accelerators.
- **Infrared radiation:** Electromagnetic radiation of longer wavelength than visible light.
- **Ion:** An atomic particle that is electrically charged, either negatively or positively.
- **Ionizing radiation:** Radiation that is capable of producing ions either directly or indirectly.
- **Irradiate:** To expose to some form of radiation.
- **Isomer:** Nuclides with the same number of neutrons and protons in different states of excitation.
- **Isomeric transition:** A relatively long-lived radioactive decay in which a nucleus goes from a higher to a lower energy state. The mass number and the atomic number are unchanged.
- **Isotope:** Isotopes of a given element have the same atomic number (same number of protons in their nuclei) but different mass numbers (different number of neutrons in their nuclei). 238U and 235U are isotopes of uranium.
- **K-capture:** The capture by an atom's nucleus of an electron from the innermost electron orbital (K-shell) surrounding the nucleus.
- **Kelvin (K):** Unit of temperature equal in size to the Celsius degree, but with the zero set by the absolute zero of temperature, -273.15°C. Ice freezes at 273 K, room temperature is about 293 K, and water boils at 373 K, at sea level. Human body temperature is 310 K.
- **Lepton:** A particle (such as the electron or neutrino) not subject to strong interactions.
- **Lifetime:** The mean life of a particle or radioactive nucleus. This is equivalent to the decay time.
- **Linear accelerator:** Particle accelerator laid out in a straight line.
- **Mass energy:** Energy a particle has by virtue of its mass (given by m times c2).
- **Mass number:** The total number of protons and neutrons in the nucleus: A=Z+N. This is also the total nucleon number of the nucleus.

- **Meson:** A particle (such as the pion) made of quark-antiquark pairs.
- **Mev:** One million electron-volts.
- **Microwaves:** Electromagnetic radiation with wavelength intermediate between radio wave and infrared radiation.
- **Multiwire proportional counter:** Particle detector using changes in the current in wires due to the passage of ionizing particles nearby.
- **Muon:** A charged lepton about 200 times more massive than an electron.
- **Muon number:** Additive quantum number characterizing muons and muon neutrinos.
- **Neutrino:** An electrically neutral particle with negligible mass. It is produced in processes such as beta decay and reactions that involve the weak force.
- **Neutron:** One of the basic particles that make up a nucleus. A neutron and a proton have about the same mass, but the neutron has no electrical charge.
- **Neutron number:** The total number of neutrons in the nucleus, N.
- **Nuclear binding energy:** The energy that free nucleons give up in order to be bound inside a nucleus.
- **Nuclear reactor:** A device in which a fission chain reaction can be initiated, maintained, and controlled. Its essential components are fissionable fuel, moderator, shielding, control rods, and coolant.
- **Nucleon:** A constituent of the nucleus; that is, a proton or a neutron.
- **Nucleus:** The core of the atom, where most of its mass and all of its positive charge is concentrated. Except for 1H, the nucleus consists of a combination of protons and neutrons.
- **Nuclide:** Any species of atom that exists for a measurable length of time. A nuclide can be distinguished by its atomic mass, atomic number, and energy state.
- **Parent:** A radionuclide that decays to another nuclide.
- **Photon:** A packet of electromagnetic energy. Photons have momentum and energy, but no rest mass or electrical charge.
- **Photomultiplier:** Commonly used device for detecting photons by converting them to an electrical signal.
- **Pion:** The least massive known spin-0 meson. The three charge states of the pion (negative, neutral and positive) are involved in the long-range force between the nucleons.
- **Proton:** One of the basic particles that makes up an atom. The proton is found in the nucleus and has a positive electrical charge equal to the

negative charge of an electron and a mass similar to that of a neutron: a hydrogen nucleus.

- **Proton number:** The total number of protons in the nucleus, Z.
- **QCD:** Quantum chromodynamics, the gauge theory describing the color strong interaction.
- **QED:** Quantum electrodynamics, the gauge theory describing electromagnetism.
- **Quark:** A strongly-interacting fermion that is a building block of hadronic matter. Quarks come in six flavors: up, down, charm, strange, top, bottom.
- **Rad (Radiation Absorbed Dose):** A former unit of an absorbed dose of ionizing radiation. One rad is equal to the absorption of radiation energy per gram of matter. It has been replaced by the gray (see above).
- **Radioactive dating:** A technique for estimating the age of an object by measuring the amounts of various radioisotopes in it.
- **Radioactive waste:** Materials that are radioactive and for which there is no further use.
- **Radioactivity:** The spontaneous decay or disintegration of an unstable atomic nucleus accompanied by the emission of radiation.
- **Radioisotope:** A radioactive isotope. A common term for a radionuclide.
- **Radionuclide:** A radioactive nuclide. An unstable isotope of an element that decays or disintegrates spontaneously, emitting radiation.
- **Rem (röntgen equivalent, man):** A measure of dose deposited in body tissue, averaged over the body. One rem is approximately the dose from any radiation corresponding to exposure to one röntgen of g radiation. The rem is no longer accepted for use with the International System. One rem is equivalent to 0.01 sievert.
- **Residual strong force:** Force between composite objects (made of quarks) due to the remaining effect of the color force on colorless objects. These forces are much weaker than the strong color force.
- **Röntgen or roentgen (R):** Unit of exposure measuring the ionizing ability of g radiation; one röntgen produces one electric charge (1.6 × 10-19° C) per 106 m^3 of dry air at 0° C and atmospheric pressure. This corresponds to an energy loss of 0.0877 joule per kilogram in air. The röntgen is no longer accepted for use with the International System.
- Scaler: An electronic instrument for counting radiation induced pulses from radiation detectors such as a Geiger-Müller tube.

- **Scintillation counter:** An instrument that detects and measures gamma radiation by counting the light flashes (scintillations) induced by the radiation.
- **Secular equilibrium:** A state of parent-daughter equilibrium that is achieved when the half-life of the parent is much longer than the half-life of the daughter. In this case, if the two are not separated, the daughter will eventually decay at the same rate at which it is being produced. At this point, both parent and daughter will decay at the same rate until the parent is essentially exhausted.
- **Shielding:** A protective barrier, usually a dense material, that reduces the passage of radiation from radioactive materials to the surroundings by absorbing it.
- **Source:** A radioactive material that produces radiation for experimental or industrial use.
- Standard Model: Gauge theory encompassing the electroweak and strong interactions.
- **Strong interaction:** The interaction due to exchange of color. Also called strong force.
- **Symmetry:** Invariance of equations of motion under changes in condition.
- **Thermal energy:** Random kinetic energy possessed by objects in a material at finite temperature.
- **Tracer:** A small amount of radioactive isotope introduced into a system in order to follow the behavior of some component of that system.
- **Transmutation:** The transformation of one element into another by a nuclear reaction.
- **Ultraviolet radiation:** Electromagnetic radiation having wavelengths between the visible part of the spectrum and X-rays.
- **Interaction:** The interaction responsible for weak decays of particles, mediated by the exchange of W± and Z0 gauge bosons.
- **X-radiation:** Electromagnetic radiation usually produced in transitions of the inner electrons of atoms. The wavelength is between ultraviolet and gamma-rays.
- **X-ray:** Electromagnetic radiation with wavelengths between ultraviolet and gamma-rays.

References

Krane, Kenneth S., and David Halliday. 1988. Introductory Nuclear Physics. New York: Wiley. ISBN 047180553X.

Martin, Brian. 2006. Nuclear and Particle Physics: An Introduction. Hoboken, NJ: Wiley. ISBN 0470025328.

Poenaru, D. N. 1996. Nuclear Decay Modes. Fundamental and Applied Nuclear Physics Series. Philadelphia: Institute of Physics. ISBN 0750303387.

Tipler, Paul, and Ralph Llewellyn. 2002. Modern Physics. 4th ed. New York, NY: W.H. Freeman. ISBN 0716743450.

Turner, James E. 1995. Atoms, Radiation, and Radiation Protection. 2nd ed. New York: Wiley. ISBN 0471595810.

4

Glimpse of Acoustic Radiation

4.1 Introduction

Acoustics is the interdisciplinary science which deals with study of mechanical waves in gases, liquids and solids including vibration, sound, ultrasound, and infra-sound. A scientist who works in the field of acoustics is an acoustician, while someone working in the field of acoustical engineering may be called an acoustical engineer. Audio engineer is concerned with the recording, manipulation, mixing, and reproduction of sound.

Applications of acoustics are present in every aspects in modern society, sub-disciplines including aero acoustics, audio signal processing, architectural acoustics, bioacoustics, electro-acoustic, environmental noise, musical acoustics, noise control, psychoacoustics, speech, ultrasound, underwater acoustics, and vibration.

Acoustic radiation directly applied to a CIED can lead to damage to the electronic circuits. Although rare, this can be encountered in the setting of extracorporeal shock wave lithotripsy, which provides focused continuous waves of energy that are used to fracture kidney and gallbladder stones.

4.1.1 Sound

Sound is an oscillation in pressure, stress, particle displacement, particle velocity, *etc.* that propagates in a medium with internal forces, or superposition of propagated oscillation. Also it is an auditory sensation which is evoked by oscillation as described above. Sound can be viewed as a wave motion in air or other elastic media where sound is a stimulus and can be further seen as excitation of hearing mechanism that results in the perception of sound. In this case, sound is a sensation.

Sound is transmitted through gases, plasma, and liquids as longitudinal waves, also called compression waves. It requires a medium to propagate. Through solids, however, it can be transmitted as both longitudinal waves and transverse waves. Longitudinal sound waves are waves of alternating pressure deviations from the equilibrium pressure, causing local regions of compression and rarefaction, while transverse waves (in solids) are waves of alternating shear stress at right angle to the direction of propagation.

The sensation of sound is a result of vibrations in the air caused by pressure fluctuations, which can be measured with a sound level meter. Vibrations are more commonly associated with mechanical systems, but air also has mass and stiffness, and as sound travels in the air it is locally compressed and expanded. The simplest type of vibration is a pure tone, which is a one-dimensional sinusoidal vibration with only one frequency component. These sinusoidal vibrations are called simple harmonic vibrations. Sound is measured with a sound level meter and can be presented by both sound pressure, measured in Pascal (Pa), and sound intensity, measured in Watt/m^2.

Amplitude

Amplitude of a periodic variable is defined as the maximum difference of the variable's extreme values over one single period, usually the distance between the top and the bottom of the curve. A large amplitude signifies high sound levels (noise) and a low amplitude signifies low sound levels (quiet). The sound wavelength is not altered when changing the amplitude, it only increases or decreases the loudness of the sound.

Frequency

The measurement unit of frequency is Hertz (Hz), and describes the number of cycles the period is repeated per second (1 Hz = 1 cycle per second). In sound theory the most used frequencies are 20Hz to 20 000 Hz (20kHz), which is the range where the human ear can respond. However, the field of acoustics also includes ultrasonic frequencies above 20kHz, and infrasonic frequencies below 20Hz.

The sound wavelength depends on the phase velocity and frequency, and frequency is further defined by the period. The period is measured in seconds and defined as the time it takes to repeat one cycle. In other words, the wavelength characterizes the distance it takes to repeat one cycle (Figure 4.1).

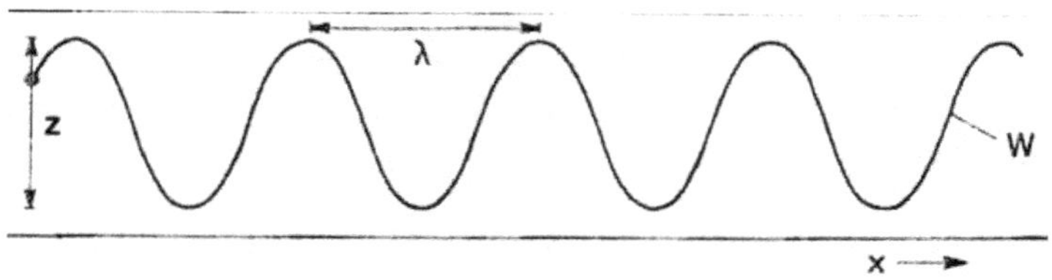

Figure 4.1: Frequency.

Typical low frequency sound has long wavelengths and contains high levels of energy, while high frequency sound has short wavelengths and contains low levels of energy. Typical low frequency sound is represented by the bass levels and high frequency sound is usually the high pitched details.

Human Ear

The human ear can respond to frequencies from 20Hz to 20kHz, but has a peak of sensitivity around 3-4kHz. This is due to the resonance of the ear cavity that amplifies certain frequencies. Many years of evolution has taught the ear to be more sensitive around the frequency spectra of speech, because that is the main source of communication between humans. Therefore the ear is not equally sensitive at all frequencies.

The measuring unit for sound is Decibel (dB), and is a logarithmic unit used to express power or intensity. The human ear can hear from 0dB to 120dB, the last being the threshold of pain. A level of 120dB or higher can be dangerous for the ear, and in worst cases contribute to permanent reduced hearing.

Because of the ear's response to different frequencies, many sound filters have been introduced in the field of acoustics. The most common one is the A-weighted filter, which is directly based upon the loudness perceived by the human ear. This means that the sound is filtered by damping the low frequency sound because the ear is less sensitive at low audio frequencies. Additionally there exists B-, C-, D- and Z-weighted filter as well. The ones which are mostly used is C-weighted and Z-weighted filters. C-weighting includes more of the low frequency sound than A-weighting, while Z-weighting is a linear flat filter.

Noise

Another type of sound is noise, which normally is undesired. The definition of noise is highly subjective, due to people reacting differently to what noise is. One may characterize noise from being casually disturbing to health damaging. Being exposed to high noise levels on a daily basis can reduce your hearing if no precautions are taken. If the ear has been critically damaged, there is little chance it will heal to its original state, once the ear has been injured there is no going back. Even though the definition of noise is very individual, acoustic engineers have a more objective way of characterizing it by using sound level meters.

4.2 Absorption and Amplification

Just as light is a wave motion that is considered as composed of particles called photons, we can think of the normal modes of vibration in a solid as being particle-like. The quantum of lattice vibration is called phonon. In lattice dynamics, we want to find the normal modes of vibration of a crystal. In other words, we need to calculate the energies (frequencies) of the phonons as a function of their wave vector's k . The relationship between frequency ω and wave vector k is called phonon dispersion.

Light and sound are similar in various ways. They both can be thought of in terms of waves, and they both come in quantum mechanical units. In the case of light we have photons, while in sound we have phonons. Both sound and light can be produced as random collections of quanta (*e.g.* light emitted by a light bulb) or orderly waves that travel in a coordinated form (*e.g.* laser light). This parallelism implies that lasers should be as feasible with sound as they are with light. In the 21st century, it is easy to produce low frequency sound in the range that humans can hear (~20 kHz), in either a random or orderly form. However, at the terahertz frequencies in the regime of phonon laser applications, more difficulties arise. The problem stems from the fact that sound travels much slower than light. This means that the wavelength of sound is much shorter than light at a given frequency. Instead of resulting in orderly, coherent phonon laser structures that can produce terahertz sound, tend to emit phonons randomly.

Acoustic radiation (sound waves) can be emitted by using the process of sound amplification based on stimulated emission of phonons. Sound (or lattice vibration) can be described by a phonon just as light can be considered as photons, and therefore one can state that SASER is the acoustic analogue of the laser.

Absorption is the transfer of the energy of a wave to matter as the wave passes through it. The energy of an acoustic, electromagnetic, or other wave is proportional to the square of its amplitude—*i.e.*, the maximum displacement or movement of a point on the wave; and, as the wave passes through a substance, its amplitude steadily decreases. If there is only a small fractional absorption of energy, the medium is said to be transparent to that particular radiation, but, if all the energy is lost, the medium is said to be opaque. All known transparent substances show absorption to some extent. For instance, the ocean appears to be transparent to sunlight near the surface, but it becomes opaque with depth.

Substances are selectively absorbing—that is, they absorb radiation of specific wavelengths. Green glass is transparent to green light but opaque to blue and red; hard rubber is transparent to infrared and X-rays but opaque to visible light. Thus, radiation of an unwanted wavelength may be removed from a mixture of waves by letting them pass through an appropriate medium. Those substances that are designed to absorb a particular wavelength or band of wavelengths are called filters.

As radiation passes through matter, it is absorbed to an extent depending on the nature of the substance and its thickness. A homogeneous substance of a given thickness may be thought of as consisting of a number of equally thin layers. Each layer will absorb the same fraction of the energy that reaches it. It is shown that a beam of waves passing from right to left through a series of layers (d1, d2, d3) of a medium. If the fractional absorption is taken as 33 percent, or 1/3, after the beam passes through the first layer d1, its initial energy (E0) will be reduced to E0/3. One-third the energy E0/3 will be absorbed passing through layer d2, and the beam will enter layer d3 with energy 1/3 (E0/3), or (E0/9). Similarly, each successive layer absorbs one-third of the energy it receives. Thus for radiation of a given wavelength, an infinitesimally thin layer will reduce the energy of a wave by a fractional amount that is proportional to the thickness of the layer. The change in energy as the wave passes through a layer is a constant of the material for a given wavelength and is called its absorption coefficient.

4.2.2 Amplification

Amplification is used to increase the size of the electrical pulses coming from the transducer after an echo is received. The amount of amplification is determined by the gain setting. The principal control associated with the amplifier is the time gain compensation (TGC), which allows the user to adjust the gain in relationship to the depth of echo sites within the body.

4.3 Sound Waves

Sound waves are mechanical longitudinal (compression) waves. A disturbance of a source (such as vibrating objects) creates an initial region of compression or high pressure. When the source vibrates, alternating regions of low and high pressure are produced in the surrounding air, called rarefactions and compressions respectively. The alternating pressure propagates through a medium, away from the source, before reaching our ears. Sound waves can be analyzed in form of pitch, duration, loudness, timbre, sonic texture and spatial location.

Pitch

Pitch is perceived as how low or high a sound is and represents the cyclic, repetitive nature of the vibrations that make up sound. For simple sounds, pitch relates to the frequency of the slowest vibration in the sound (fundamental harmonic). In the case of complex sounds, pitch perception can vary. Sometimes individuals identify different pitches for the same sound, based on their personal experience of particular sound patterns. Selection of a particular pitch is determined by pre-conscious examination of vibrations, including their frequencies and the balance between them. Specific attention is given to recognizing potential harmonics. Every sound is placed on a pitch continuum from low to high. For example: white noise (random noise spread evenly across all frequencies) sounds higher in pitch than pink noise (random noise spread evenly across octaves) as white noise has more high frequency content. During the listening process, each sound is analyzed for a repeating pattern and the results forwarded to the auditory cortex as a single pitch of a certain height and chroma.

Duration

Duration is perceived as how "long" or "short" a sound is, and relates to the onset and offset signals created by nerve responses to sounds. The duration of a sound usually lasts from the time the sound is first noticed until the sound is identified as having changed or ceased. Sometimes this is not directly related to the physical duration of a sound. In noisy environment, gapped sounds when they are continuous because the offset messages are missed owing to disruptions from noises in the same general bandwidth. This can be of great benefit in understanding distorted messages such as radio signals that suffer from interference, as the message is heard as if it was continuous.

Loudness

Loudness is perceived as how "loud" or "soft" a sound is and relates to the totaled number of auditory nerve stimulations over short cyclic time periods, most likely over the duration of theta wave cycles. It shows that at short durations, a very short sound can sound softer than a longer sound even though they are presented at the same intensity level. Past around 200 ms, this is no longer the case, and the duration of the sound no longer affects the apparent loudness of the sound. Figure 3 gives an impression of how loudness information is summed over a period of about 200 ms before being sent to the auditory cortex. Louder signals create a greater 'push' on the Basilar membrane and thus stimulate more nerves, creating a stronger loudness signal. A more complex signal also creates more nerve firings and so sounds louder (for the same wave amplitude) than a simpler sound, such as a sine wave.

Timbre

Timbre is perceived as the quality of different sounds and represents the pre-conscious allocation of a sonic identity to a sound. This identity is based on information gained from frequency transients, noisiness, unsteadiness, perceived pitch and the spread and intensity of overtones in the sound over an extended time frame. The way a sound changes over time provides most of the information for timbre identification. Even though a small section of the wave form from each instrument looks very similar, differences in changes over time between the clarinet and the piano are evident in both loudness and harmonic content. Less noticeable are the different noises heard, such as air hisses for the clarinet and hammer strikes for the piano.

Sonic Texture

Sonic texture relates to the number of sound sources and the interaction between them. The word 'texture', in this context, relates to the cognitive separation of auditory objects. In music, texture is often referred to as the difference between unison, polyphony and homophony, but it can also relate (for example) to a busy cafe; a sound which might be referred to as 'cacophony'. However texture refers to more than this. The texture of an orchestral piece is very different from the texture of a brass quintet because of the different numbers of players. The texture of a market place is very different from a school hall because of the differences in the various sound sources.

Spatial Location

It shows cognitive placement of a sound in an environmental context; including the placement of a sound on both the horizontal and vertical plane, the distance from the sound source and the characteristics of the sonic environment. In a thick texture, it is possible to identify multiple sound sources using a combination of spatial location and timbre identification. It is the main reason why we can pick the sound of an oboe in an orchestra and the words of a single person at a cocktail party.

Noise

Noise is a term often used to refer to an unwanted sound. In science and engineering, noise is an undesirable component that obscures a wanted signal. However, in sound perception it can often be used to identify the source of a sound and is an important component of timbre perception.

Soundscape

Soundscape is the component of the acoustic environment that can be perceived by humans. The acoustic environment is the combination of all sounds (whether audible to humans or not) within a given area as modified by the environment and understood by people, in context of the surrounding environment.

4.4. Sources of Sound

4.4.1 Radiation from a Monopole Source

A monopole is a source which radiates sound equally well in all directions. The simplest example of a monopole source would be a sphere whose radius alternately expands and contracts sinusoidally. The monopole source creates a sound wave by alternately introducing and removing fluid into the surrounding area. A boxed loudspeaker at low frequencies acts as a monopole. The directivity pattern for a monopole source is shown (Figure 4.2).

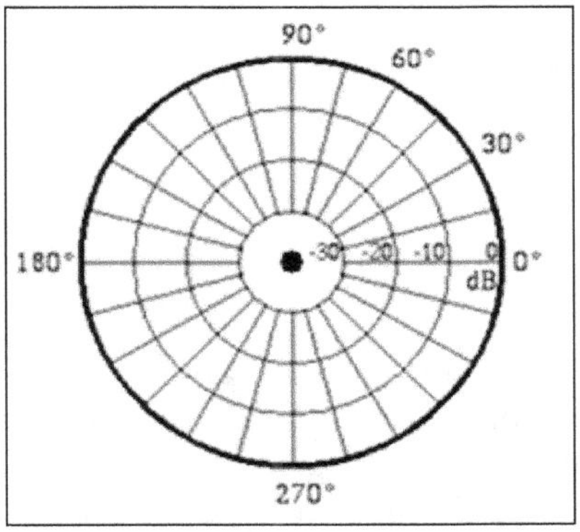

Figure 4.2: Directivity Pattern for a Monopole Source.

The figure below shows the pressure field produced by a monopole source (Figure 4.3). Individual points on the grid simply move back and forth about some equilibrium position while the spherical wave expands outwards.

Figure 4.3: Pressure Field Produced by a Monopole Source.

4.4.2 Radiation from a Dipole Source

A dipole source consists of two monopole sources of equal strengths but opposite phase, and separated by a small distance compared with the wavelength of sound. While one source expands the other source contracts. The result is that the fluid (air) near the two sources sloshes back and forth to produce the sound. A sphere which oscillates back and forth acts like a dipole source, as does an unboxed loudspeaker (while the front is pushing outwards the back is sucking in). A dipole source does not radiate sound in all directions equally. The directivity pattern shown at right looks like the Figure 4.4, where, there are two regions where sound is radiated very well, and two regions where sound cancels.

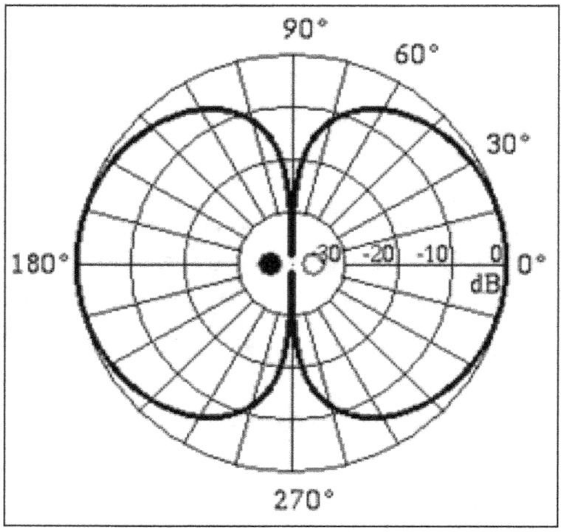

Figure 4.4: Sound Radiating in Two Regions.

The Figure 4.5 shows the pressure field produced by a dipole source. At the center of the pressure field you can see sloshing back and forth caused by the dipole motion. The regions where sound is cancelled shows up along the vertical axes (the grid motion is almost zero). Furthermore, the wavefronts expanding to the right and left are 180o out of phase with each other.

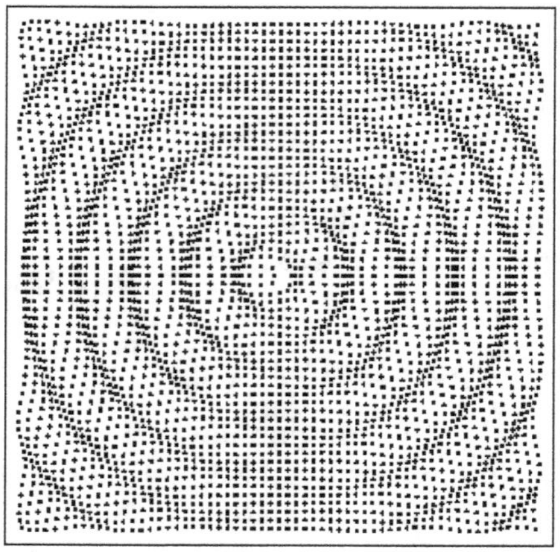

Figure 4.5: Pressure Field by Dipole Source.

4.4.3 Radiation from a Lateral Quadrupole Source

If two opposite phase monopoles make up a dipole, then two opposite dipoles make up a quadrupole source. In a Lateral Quadrupole arrangement the two dipoles do not lie along the same line (four monopoles with alternating phase at the corners of a square). The directivity pattern for a lateral quadrupole looks like a clover-leaf pattern; sound is radiated well in front of each monopole source, but sound is canceled at points equidistant from adjacent opposite monopoles.

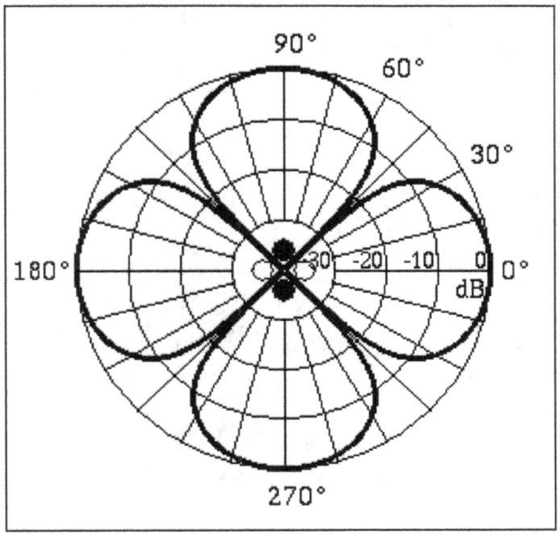

Figure 4.6: Quadrupole Arrangement the Two Dipoles.

The Figure 4.7 shows the pressure field produced by a lateral quadrupole source. At the center of the pressure field you can see the quadrupole motion as the particles alternate motion in the horizontal and vertical directions back and forth caused by the dipole motion. The regions where sound is cancelled shows up along the diagonals (where the grid motion is almost zero). Furthermore, there is a 180o phase difference between the horizontal and vertical wavefronts.

Figure 4.7: Field Produced by Lateral Quadrupole Source.

4.4.4 Radiation from a Linear Quadrupole Source

If two opposite phase dipoles lie along the same line they make up a Linear Quadrupole source. A tuning fork is a good example of a linear quadrupole source (each tine acts as a dipole as it vibrates back and forth, and the two tines oscillate in opposite directions). What makes the linear quadrupole interesting is that there is a very obvious transition from near field to far field. In the near field there are four maxima and four minima, with the maxima along the quadrupole axis about 5dB louder than the maxima perpendicular to the quadrupole axis. The near field directivity pattern is shown in Figure 4.8.

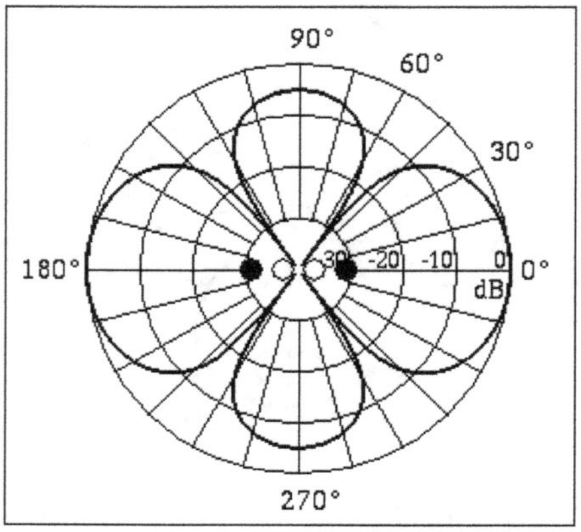

Figure 4.8: Near Field Directivity Pattern.

In the far field there are only two maxima (along the quadrupole axis) and two minima (perpendicular to the quadrupole axis) as shown in the Figure 4.9.

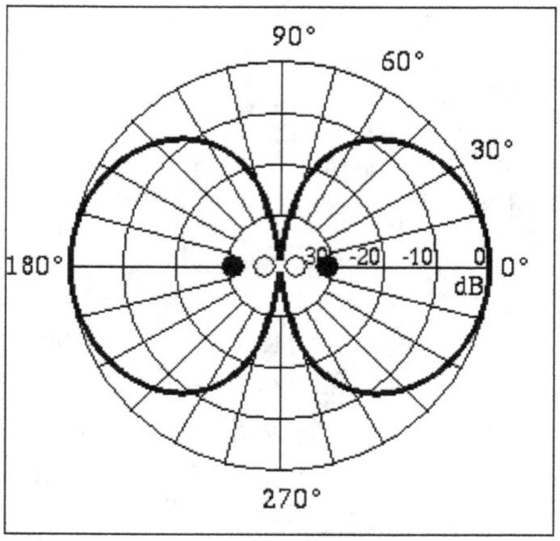

Figure 4.9: Far Field Directivity Pattern.

The Figure 4.10 shows the pressure field radiated by a linear quadrupole. At the center of the picture you can see the quadrupole near field pattern. As the wave expands outwards it becomes almost a spherical wave (notice that the left and right going wavefronts are in phase as opposed to the dipole source) except that the amplitude is severely reduced in the vertical directions.

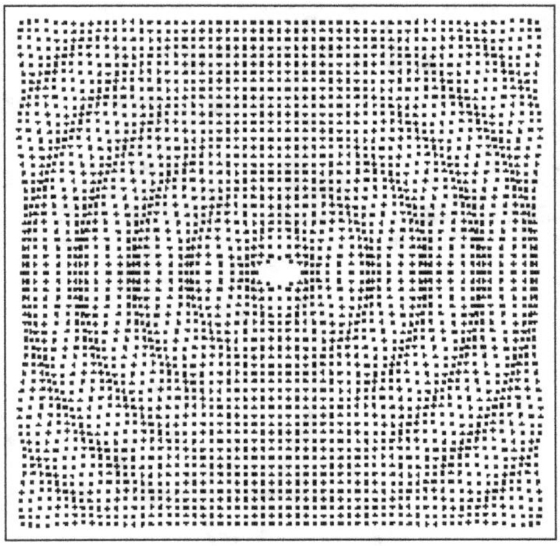

Figure 4.10: Pressure Field Radiated by Linear Quadrupole.

4.5 Sound Pressure and Vibrations

Sound intensity and sound pressure can be measured directly by suitable instrumentation. Sound power can be calculated from measured values of sound pressure or sound intensity levels and a knowledge of the area over which the measurements were made. The main use of sound power is for the noise rating of machines *etc.*, and sound intensity is mainly used for location and rating of noise sources. When it comes to evaluation of the harmfulness and annoyance of noise sources, sound pressure is the important parameter.

4.5.1 Sound Pressure

Since a sound wave consists of a repeating pattern of high-pressure and low-pressure regions moving through a medium, it is sometimes referred to as a pressure wave. If a detector, whether it is the human ear or a man-made instrument, were used to detect a sound wave, it would detect fluctuations in pressure as the sound wave impinges upon the detecting device. At one instant in time, the detector would detect a high pressure; this would correspond to the arrival of a compression at the detector site. At the next instant in time, the detector might detect normal pressure.

Further a low pressure would be detected, corresponding to the arrival of a rarefaction at the detector site. The fluctuations in pressure as detected by the detector occur at periodic and regular time intervals. In fact, a plot of pressure versus time would appear as a sine curve. The peak points of the sine curve correspond to compressions; the low points correspond to rarefactions; and the "zero points" correspond to the pressure that the air would have if there were no disturbance moving through it. The Figure 4.11 shows correspondence between the longitudinal nature of a sound wave in air and the pressure-time fluctuations that it creates at a fixed detector location.

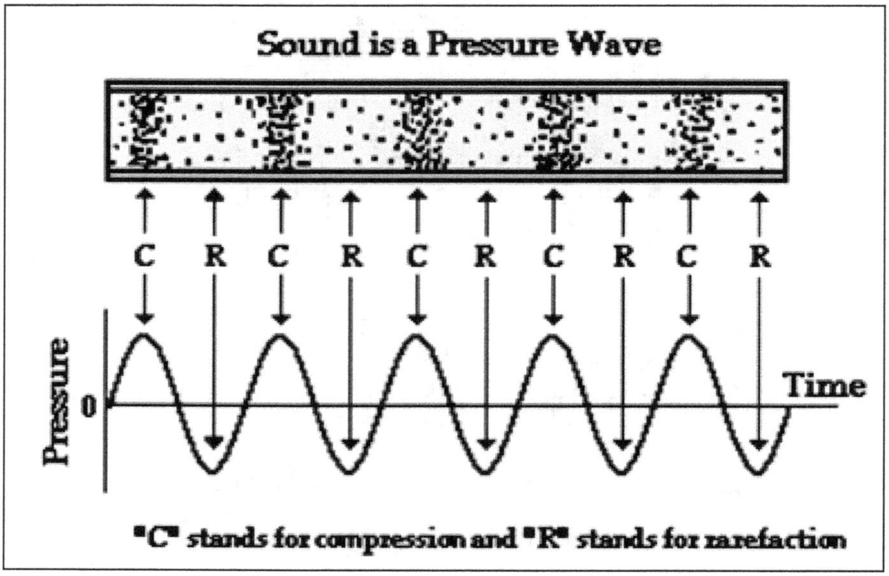

Figure 4.11: Longitudinal Nature of Sound Wave in Air with Pressure-Time Fluctuations.

When a sound source such as a tuning fork vibrates it sets up pressure variations in the surrounding air. The emission of the pressure variations can be compared to the ripples in a pond caused by a stone thrown in the water.

The ripples spread out from the point where the stone entered. However the water itself does not move away from the center. The water stays where it is, moving up and down to produce the circular ripples on the surface. Sound is like this. The stone is the source, the pond is the air, and the ripples are the resulting sound wave.

dB - Decibel

The advantage of using dB's is clearly seen when a dB scale is drawn on the illustration shown earlier. The linear scale with its large and unwieldy numbers is converted into a much more manageable scale from 0 dB at the threshold of hearing (20 mPa) to 130 dB at the threshold of pain.

Lp = 20 log (p/p0) dB re 20 mPa

(p0 = 20 mPa = 20 × 10-6 Pa)

The sound pressure level, Lp, in dB's is defined as 20 log p/p0 , where p is the measured value in Pa, and p0 is a standardized reference level of 20 mPa - the threshold of hearing. Note here that the word level is added to sound pressure to indicate that the quantity has a certain level above the reference level, and the symbol for sound pressure level is Lp. Here it shows two examples of how to use the formula for calculation of dB levels. Besides being examples of calculation, these two levels are interesting because they are the levels used for sound level meter calibration

Sound Fields

The source we looked at earlier is called a point source and for such a source it was mentioned that the sound pressure drops to half its value when the distance to the source is doubled. This corresponds to a drop in sound pressure of 6 dB.

Another type of source is the line source, which could be a pipe carrying a turbulent fluid, or a road with a high traffic flow. The sound pressure from a line source only drops by approximately 3 dB for a doubling of distance from the source, because the sound spreads out from the source as a wavefront in a direction perpendicular to the line source.

The most rarely found type of sound source when dealing with normal noise measurements is the plane source. A plane source will in principle consist of a piston from which energy is radiated into a tube setting up a plane wave in the tube. Assuming no loss of energy through the walls of the tube, the intensity, *i.e.* the acoustic energy flowing through the tube, is independent of the distance from the source. Since the intensity is the same everywhere in the tube, the sound pressure level will not drop with an increase in distance from the piston.

In practice, the majority of sound measurements are made in rooms that are neither anechoic nor reverberant - but somewhere in between. This makes it difficult to find the correct measuring positions where the noise emission from a given source must be measured.

It is normal practice to divide the area around a noise source *e.g.* a machine into four different fields:

- Near field
- Far field
- Free field
- Reverberant field

The near field is the area very close to the machine where the sound pressure level may vary significantly with a small change in position. The area extends to a distance less than the wavelength of the lowest frequency emitted from the machine, or at less than twice the greatest dimension of the machine, whichever distance is the greater. Sound pressure measurements in this region should be avoided. The far field is divided into the free field and the reverberant field.

In the free field the sound behaves as if in open air without reflecting surfaces to interfere with its propagation. This means, that in this region the sound level drops 6 dB for a doubling in distance from the source.

Table 4.1: Sound Levels L and Corresponding Sound Pressure and Sound Intensity

Examples	Sound Pressure Level L_p dB SPL	Sound Pressurep N/m² = Pa	Sound IntensityI W/m²
Jet aircraft, 50 m away	140	200	100
Threshold of pain	130	63.2	10
Threshold of discomfort	120	20	1
Chainsaw, 1m distance	110	6.3	0.1
Disco, 1 m from speaker	100	2	0.01
Diesel truck, 10 m away	90	0.63	0.001
Kerbside of busy road, 5 m	80	0.2	0.0001
Vacuum cleaner, distance 1 m	70	0.063	0.00001
Conversational speech, 1m	60	0.02	0.000001
Average home	50	0.0063	0.0000001
Quiet library	40	0.002	0.00000001
Quiet bedroom at night	30	0.00063	0.000000001
Background in TV studio	20	0.0002	0.0000000001
Rustling leaf	10	0.000063	0.00000000001
Threshold of hearing	0	0.00002	0.000000000001

Example Sound Pressure Level for two persons talking (not shouting) at various distances (level at the receiver's ear)

0.25m	0.5m	1m	1.5m	2m	3m
70-76dB	65-71dB	58-64dB	55-61dB	52-58dB	50-56dB

4.5.2 Sound Vibrations

Vibrations are small oscillation disturbances of the particles in a given body, such as water or a string. Regular vibrations have a defined period, the amount of time it takes to complete a cycle, while irregular vibrations do not, like those created by snare drums or a giant wave crashing on the ocean shore. The regular, periodic vibrations have a given frequency, f, in cycles per second.

When struck, a tuning fork vibrates at a single, particular frequency. The constant movement of the prongs back and forth causes repeated sound impulses, which are really disturbances in the air (6). The prongs hit the air, and the air continues to hit neighboring air molecules. At one instant, some molecules are close while others are further apart, a phenomenon known as condensation and rarefaction of air as shown in Figure 4.12.

Such pattern of propagation is a wave, and in one vibration, the molecules move one wavelength, λ. After measuring the frequency, the sound wave's velocity can be given as $V=f\lambda$ (1). The more common, sinusoidal depiction of a given wave comes from an oscilloscope, an apparatus that senses the pressure from a wave and translates it into an electric signal. Mathematically modeling a wave's propagation requires a second order partial differential equation.

Instruments produce a fundamental tone, the most audible tone, as well as many additional frequencies above the pitch, known as overtones. The fundamental and the additional overtones form the series of harmonics that give a sound or pitch a certain quality. A note's unique quality, or timbre, is based on the relative energies of the harmonics. In other words, each note sounds the way it does because the wave is a complex combination of frequencies.

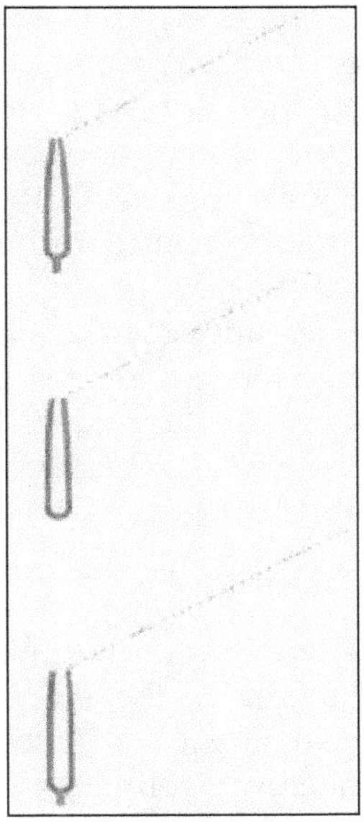

Figure 4.12: Tuning Fork Vibration.

Vibrating Strings

Musical instruments create sounds through the physical communication of a primary vibrator, resonant vibrator, and a sound effuser. These components provide the initial vibrations, amplify the vibrations, and allow the vibrations to escape. String instruments like the viola or cello amplify the sounds made by bowing or plucking the strings. The strings are held tightly around the pegs at one end and the tailpiece at the other end. When the string vibrates, the vibrations propagate down to the bridge, which carries them to the soundboard that spans the inside of the wooden body. The soundboard amplifies the vibration, and the sound waves emerge through the two f-shaped holes. Marin Mersenne, a sixteenth century French mathematician and the "father of acoustics," devised three laws to calculate the frequency of the fundamental tone produced by a string. Together, they state that the frequency is inversely proportional to the length, proportional to the square root of the tension, and inversely proportional to the square root of the mass per unit length, as shown in expression.

$$f = \frac{1}{2L}\sqrt{\frac{T}{M}}$$

As the string is pulled more tautly on a violin by turning the peg towards the scroll, the pitch increases, and as the musician presses on the string, the length of the string decreases and the frequency increases. Cellos and basses have longer strings than violins and violas, which is why these larger instruments can play in a lower pitch register. Consequently, a musician can produce the same pitch on different string; this causes challenges in determining optimal "fingerings" when studying a piece of music.

While string instruments typically have four or six strings, the piano has 230 (two or three per note). An enormous amount of combined tension, up to 30 tons in a concert grand piano, compensates for the great length of the strings. Since the fundamental frequencies of the piano range from 27 Hz for the lowest A to 4096 Hz for highest C, it would be farcical to make the A1 (lowest note) strings 150 times longer than the C8 (highest note) strings. Instead, both the tension and the length vary for each note. The piano is especially unique with its equal temperance. Ever since the days of Johann Sebastian Bach, the difference between each key's frequency has been equal. This allows pieces to sound pleasing in any key. As a result, the frequencies are not perfectly aligned with the whole number ratio patterns that characterize "consonance," but they are only a few hundredths of a hertz off.

Vibrating Air

The vibration of air in woodwind instruments to produce sounds comes in two categories: the edge tones of direct vibrations between the musician and the instrument, as in the flute and piccolo, and the vibration of a reed indirectly causing a sound, as in the clarinet and oboe. In both cases, the player supplies the energy to cause a vibration. The molecules are set into random motion when the player blows a note, moving the sound forward within the column. Molecules in a column of air, just as in a string, have a frequency of free vibration and can be excited by matching frequencies. Similar to string instruments, the standing waves produced have a fundamental tone and many overtones. The pitch produced by many wind and brass instruments depends on the embouchure, or shape of the mouth and tongue when creating a note, as well as the keys pressed, both of which alter the size of the column through which the air travels.

Brass instruments are set up much differently. The trombone has a slide to change the size of the air column and the further out the slide, the lower the note. Trumpets have three valves: the first lowers the frequency by a whole tone, the second lowers the frequency by a half tone, and the third raises the frequency by a whole tone. Again, the embouchure allows a trumpeter to create a wide range of notes.

4.6 Sound Transmission

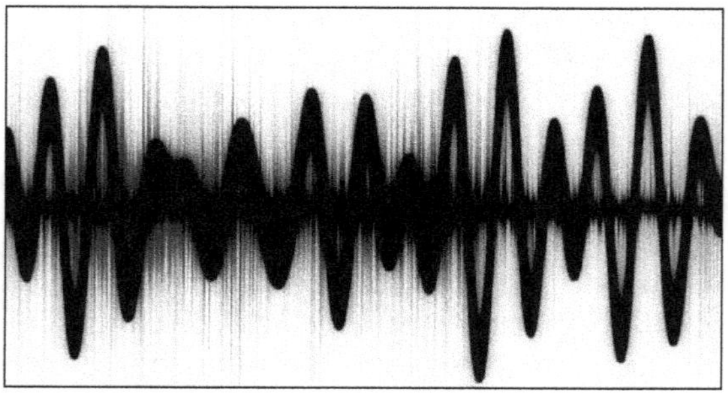

Figure 4.13: Sound Transmission.

Sound is transmitted by particles (atoms or molecules) in a solid, liquid or gas colliding with each other. It is a wave which is created by vibrating objects and propagated through a medium (solid, liquid or gas) from one location to another. A vibrating source can be a sound coming from a drum, speaker of a radio, the mouth of a person (vocal chord), a car engine, a plane above the sky and so on. Although sound is commonly associated with air, sound will readily travel through many materials that are solid, liquid and gas (Figure 4.13).

Solids are made up of particles (atoms) that do not move about because they are closely packed (touching each other) and held together by strong intermolecular forces. Therefore, they are always in a fixed position and can only vibrate in a fixed position, sending sound waves along its path very fast. This is similar to a domino effect. This means that sound waves (produced from a vibrating source) is immediately transmitted by the vibrating particle in a fixed position by hitting the neighboring atom sending on the sound wave from one atom to the other throughout the solid object. In liquid, the particles are constantly moving and so the particles change their position, but they are not fixed to each other that strongly as they are in a solid. They can also vibrate and collide with other of their particles over a short range; sending sound waves through it. This is because the attractive forces among the particles of a liquid are sufficiently strong enough to hold them together but are not strong enough to hold them in a fixed position (Figure 4.14).

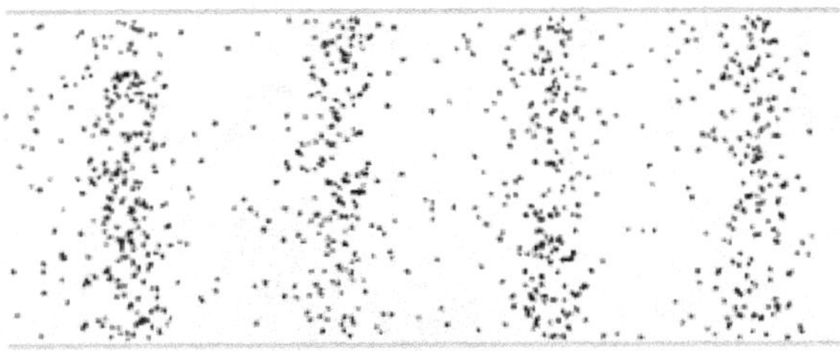

Figure 4.14: Movement of Particles.

As such, sound will travel slightly less fast through liquids. In gas, particles in gases are very far apart from each other. They can vibrate as well as move freely and randomly in all direction. This is due to the fact that the force of attraction between the particles are negligible. This means therefore, that there is no force of attraction between gas particles. Gas molecules must move quite a distance before they collide with other molecules. Sound energy cannot move as quickly when the molecules are not in contact with each other. Thus, since gas particles are very far apart, sound waves traveling through this medium, will be much slower than that of a liquid and a solid respectively. Because of gas particles being everywhere on Earth, we can hear sound from all directions.

While sound will readily travel through many materials, in order for us to hear it, it will have to travel through the air to reach our ears (eardrum). Your eardrum vibrates from sound waves to allow you to sense them. We might be able to detect it through vibration of solids, the ripple in liquid but you can only detect it in air by hearing it. In spite of this, humans cannot hear all sounds that travel through the air. It will have to be at a certain decibel for us to hear it. For *e.g.* dogs can hear sounds that cannot be detected by humans. Sound can also be detected by other means other than hearing it. Sounds traveling in any of the three states can be detected using electronic detectors *e.g.* Microphones.

Sound (in air) is made when air molecules vibrate and move (away from the vibrating source) in a pattern called waves, or sound waves. Sound is a mechanical, longitudinal wave (that moves in all directions) and travels in waves of compressions and rarefactions (expansion) as it successively passes through a medium.

Air particles as shown above will move in, to and fro position where they collide with each other to form a cluster (compression) then they rebound from each other, the same effect of hitting a ball against a wall (rarefaction) – *i.e.* a series of mechanical compressions and rarefactions. Sounds are carried this way by the air particle in an outward fashion similar to ripple in a pond as it continues to collide with air particles. The air particles do not move over a large area but, carry the sound by colliding with other air particles. This way it passes on the vibration similar to a domino effect, as it moves away from the vibrating source. Vibrations of frequencies in the approximate range of 15 to 20,000 hertz, can be detected by human ears (Figure 4.15).

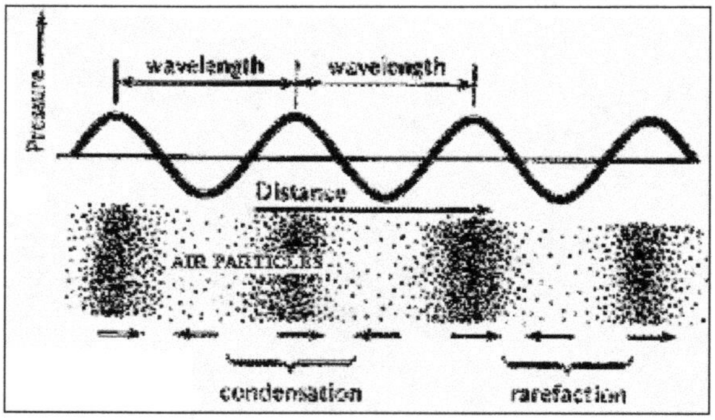

Figure 4.15: Pressure vs. Distance of Particles.

When a sound wave impacts upon the surface of a solid body, some portion of its energy will be reflected, some absorbed and the rest transmitted through the body. The relative proportion of each depends on the nature of the material impacted. This topic concentrates on the transmitted component.

If we consider the transmission of sound through a partition, we can actually measure the sound energy on both the source side (Wsrc) and the receiving side (Wrec) to determine exactly what fraction of the sound is transmitted through. We can thus determine the transmission coefficient (t) for that partition as follows:

$$t = Wrec/Wsrc$$

Transmission Loss (TL) or Sound Reduction Index (SRI) are used to describe the reduction in sound level resulting from transmission through a material which are given by:

$$SRI = 10 \log (Wsrc/Wrec)$$

$$= 10 \log (1/t)$$

$$= -10 \log (t)$$

Resonant Frequencies

At slightly higher frequencies the resonance of the wall begins to control its transmission behaviour. Because every panel has a finite boundary and edge fixings, it will have a series of natural frequencies at which it will vibrate more easily than others. These are called resonant frequencies and consist of a fundamental frequency (having the greatest effect), and integer multiples of this fundamental called harmonics (having less and less effect). The fundamental resonant frequency of a panel can be calculated as follows:

Fr = 0.45 * vL * b((1/l)Ã,Â² + (1/h)Ã,Â²)

vl = sqrt(E/(p * (1 – sÃ,Â²)))

where:

b = the panel thickness (m),

l and h = length and height(m), and

vl = the longitudinal velocity of sound in the partition (m/s).

In the calculation of vl:

E = Young's modulus of elasticity,

s = it's Poisson ratio, and

p = density (kg/mÃ,Â³)

Glossary

- **Acoustics:** the science of sound. Its production, transmission and effects. The branch of physics that treats the phenomena and laws of sounds as it effects people.
- Airborne sound: sound that reaches the point of interest by propagation through air.
- **Articulation index (ai):** a measure of speech intelligibility influenced by acoustical environment rated from 0.01 to 1.00.the higher the number the higher the intelligibility of words and sentences understood from 0-100%.
- **Absorption:** the properties of a material composition to convert sound energy into heat thereby reducing the amount of energy that can be reflected.
- **Attenuation:** the reduction of sound energy as a function of distance traveled. (see also inverse square law).
- **Background noise:** the sum total of all noise generated from all direct and reflected sound sources in a space that can represent an interface to good listening and speech intelligibility. (hearing impaired persons are especially victimized by background noise).
- **Barrier:** anything physical or an environment that interferes with communication or listening. A poor acoustical environment can be a

barrier to good listening and especially so for persons with a hearing impairment.

- **Cycle:** in acoustics, the cycle is the complete oscillation of pressure above and below the atmospheric static pressure.
- **Decibel (db):** sound level in decibels as a logarithmic ratio. Sound intensity described in decibels. *I.e.*:
- **Deflection:** the distance an elastic body or spring moves when subjected to a static or dynamic force. Typical units are inches or mm.
- **Diffusion:** the scattering or random reflection of a sound wave from a surface. The directions of reflected sound is changed so that listeners may have sensation of sound coming from all directions at equal levels.
- **Ear:** an incredible hearing mechanism consisting of outer, middle and inner ear segments that cause sound pressures to be picked up by the ear that are transmitted through auditory nerves where signals are interpreted by brain as sound.
- **Echo:** reflected sound producing a distinct repetition of the original sound. Echo in mountains is distinct by reason of distance of travel after original signal has ceased.
- **Frequency:** the number of oscillations or cycles per unit of time. Acoustical frequency is usually expressed in units of hertz (hz) where one hz is equal to one cycle per second.
- **Hertz (hz):** frequency of sound expressed by cycles per second. (see cycle).
- **Loudness:** the average deviation above and below the static value due to sound wave is called sound pressure. The energy expended during the sound wave vibration is called intensity and is measured in intensity units. Loudness is the physical resonance to sound pressure and intensity.
- **Noise:** unwanted sound that is annoying or interferes with listening. Not all noise needs to be excessively loud to represent an annoyance or interference.
- **Noise reduction (nr):** the amount of noise that is reduced through the introduction of sound absorbing materials. The level (in decibels) of sound reduced on a logarithmic basis.
- **Pitch:** the perceived auditory sensation of sounds expressed in terms of high or low frequency stimulus of the sound.
- **Reflection:** the amount of sound wave energy (sound) that is reflected off a surface. Hard non-porous surfaces reflect more sound than soft-porous surfaces. Some sound reflection can enhance quality of signal of speech and music. (see echo).
- **Resonance:** the emphasis of sound at a particular frequency.

- **Resonant frequency:** a frequency at which resonance exists.
- **Reverberation:** the time taken for sound to decay 60 db to 1/1,000,000 of its original sound level after the sound source has stopped. Sound after it has ended will continue to reflect off surfaces until the wave loses enough energy by absorption to eventually die out. Reverberation time is the basic acoustical property of a room which depends only on its dimensions and the absorptive properties of its surfaces and contents. Reverberation has an important impact on speech intelligibility.
- **Signal to noise ratio:** is the sound level at the listener's ear of a the speaker above the background noise level. The inverse square law impacts on the s/n ratio. Signal to noise ratios are important in classrooms and should be in range of 15 to 20 db.
- **Sound:** sound is an oscillation in pressure, stress particle displacement, particle velocity in a medium – in room temperature. (in air speed of sound is 1125'/second or one mile in 5 seconds.) Sound produces an auditory sensation caused by the oscillation.
- **Sound level:** a subjective measure of sound expressed in decibels as a comparison corresponding to familiar sounds experienced in a variety of situations.
- **Sound pressure:** the total instantaneous pressure at a point in space, in the presence of a sound wave, minus the static pressure at that point.
- **Sound pressure level:** the sound pressure level, in decibels, of a sound is 20 times the logarithm to the base 10 of the ratio of the sound pressure to the reference pressure. The reference pressure shall be explicitly stated and is defined by standards.
- **Spl:sound pressure level:** quantity used to describe the loudness of a sound. The sound pressure level is expressed in decibels and is measured with a sound level meter. For example, a conversation between two people inside an average-size room will produce an average "a" weighted sound pressure level of 50 to 55 lb.
- **Spectrum:** the description of a sound wave's components of frequency and amplitude.
- **Ultrasounds**: sounds of a frequency higher than 20,000 hz. The frequency region containing these frequencies is called the ultrasonic region.
- **Vibration:** a force which oscillates about some specified reference point. Vibration is commonly expressed in terms of frequency such as cycles per second (cps), hertz (hz), cycles per minute (cpm) or (rpm) and strokes per minute (spm). This is the number of oscillations which occurs in that time period. The amplitude is the magnitude or distance of travel of the force.

- **Volume:** the cubic space of a room bounded by walls, floors, and ceilings determined by volume = length × width × height of space. Volume influences reverberation time.
- **Wavelength:** sound that passes through air it produces a wavelike motion of compression and rarefaction. Wavelength is the distance between two identical positions in the cycle or wave. Similar to ripples or waves produced by dropping two stones in water. Length of sound wave varies with frequency. Low frequency equals longer wavelengths.

References

A. S. Davydov, Quantum Mechanics, Vol. 1 (Pergamon, Oxford, 1965),

D. Pines, Elementary Excitations in Solids (Benjamin, New York, 1963).

E. G. Harris, in Advances in Plasma Physics, Vol. 3, edited by A. Simon and W. B. Thompson (Wiley-Interscience, New York, 1969).

E. M. Epshtein, JETP Lett. 13, 364 (1971).

E. M. Epshtein, Sov. Phys. Semicond. 11, 243 (1977).

O. A. C. Nunes, Solid State Commun. 42, 451 (1982).

O. A. C. Nunes, Solid State Commun. 45, 53 (1983).

5

Interaction of Charged Particles, Photons and Neutrons

5.1 Types of Charged Particles Interaction

Gamma-rays, X-rays, neutrons, and neutrinos all have no net charge - they are electrostatically neutral. In order to detect them they must interact with matter and produce an energetic charged particle. In the case of gamma and X-rays, a photo-electron is produced. In the case of neutrons, a proton is given kinetic energy in a billiard ball collision. So we begin our discussion of charged particle interactions by demonstrating that even when detecting neutral particles we must quickly think in terms of the charged particles.

Charged particles can be divided into:

- Electrons, including positrons and beta particles from radioactive decay, Auger electrons, and internal conversion electrons
- Heavy charged particles such as the alpha particle, fission fragments, protons, deuterons, tritons, and mu and pi mesons.

The charged particles interact with matter primarily through the Coulomb force. Differences in the energy deposition trails between the two types of charged particles are due to the mass differences, *i.e.*, the more easily altered trajectory of the electron. The maximum energy that can be transferred from a charged particle to an electron in a single collision is given by 4 T m(e)/m, for about 1/500 of the particle energy per nucleon for heavy particles. Due to this small fraction and the fact that at any time the particle is interacting with more than one electron, the heavy particles appear to continuously slow down, gradually losing their kinetic energy along an unaltered linear path.

When a charged particle gives sufficient energy to an orbital electron such that the freed electron itself can cause secondary ionization, we say that a delta-ray has been produced. The majority of the energy loss of a heavy charged particle such as the alpha occurs via these delta-rays.

5.1.1 Interactions

We will concern ourselves primarily with two types of interactions: collisional and radiative:

- **Collisional:** Inelastic collision with atomic electrons. This results in excitation or ionization. These processes ultimately end with the heating of the absorber (through atomic and molecular vibrations) unless the ions and electrons can be separated using an electric field as is done in radiation detectors.
- **Radiative:** Inelastic collision with nucleus. A quantum of electromagnetic radiation is emitted (a photon). Energy loss is experienced by the particle. Important for electrons. Probability of nuclear excitation is negligible. This process is also known as RADIATIVE energy loss.

5.1.2 Components

Stopping Power

The stopping of charged particles in matter is by collisional and radiative processes which occur in frequencies dictated by their interaction cross sections (or probabilities). What we observe then is a statistical average of the 2 processes occurring as the particle slows down. Linear Stopping Power is given by S= - dT/dx,

where

T = charged particle kinetic energy

–dT = energy increment lost in infinitesimal material thickness of dx

The units of stopping power are keV/micron where higher the stopping power, the shorter the range into the material the particle can penetrate. The quantity S is referred as specific energy loss. The stopping power S increases as the particle velocity is decreased. The classical depiction of the charged particle interaction with an electron is depicted in the Figure 5.1.

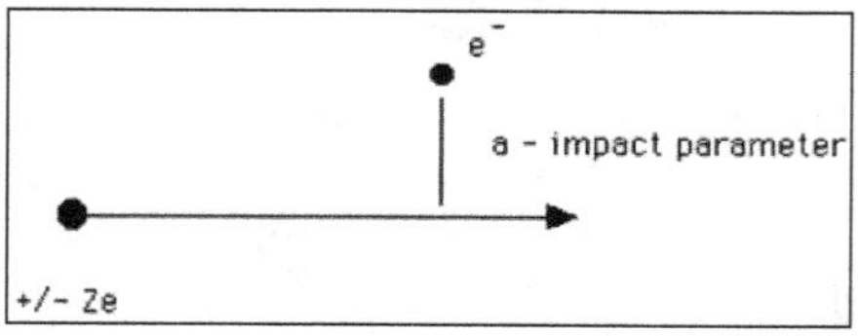

Figure 5.1: Charged Particle Interaction.

Heavy Particles

The graph shows specific energy loss along the track of charged particle known as Bragg Curve. A typical Bragg curve is depicted in the Figure 5.2 for an alpha particle of several MeV of initial energy:

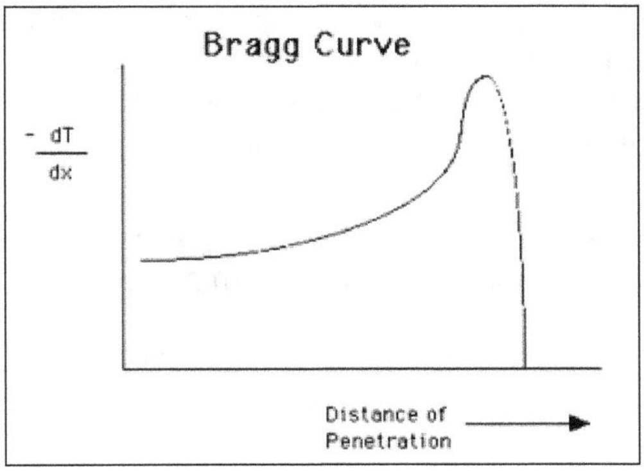

Figure 5.2: Bragg Curve.

As the energy falls, the specific energy loss increases according to the Bethe-Bloch formula. As the energy falls below a threshold, however, an electron will attach to the alpha, dramatically lowering the specific energy loss.

Bremsstrahlung and Stopping Power

The total stopping power for electrons can be given as a combination of the collisional and radiative types of interactions:

[dT/dx] total = [dT/dx]collision + [dT/dx] radiative

(for heavy particles we only consider orbital electron interactions (an atomic process) since the probability of nuclear interaction resulting in energy loss is much smaller.)

The radiative component calculated from electromagnetic and relativistic theory is given by Eqn

$$-\left(\frac{dT}{dx}\right)_r = \frac{NTZ(Z+1)e^4}{137\, m_o^2 c^4}\left(4 \ln \frac{2T}{m_o c^2} - \frac{4}{3}\right)$$

which is the basis for the approximation for the percentage of energy lost to radiation for monoenergetic electrons:

(dT/dx)r/(dT/dx)total = EZ/1000 ; where E is in MeV, where Z is the atomic number of the absorber and E is the charged particle energy in MeV.

For example, a stream of 100 keV electrons hits a tungsten target (Z=74). Only 0.74% of the energy lost goes to emitted X-rays.

For polyenergetic beta rays, above eqn becomes:

(dT/dx)r/(dT/dx)total = Z E(max)/3000

Since the average beta energy is 1/3 that of the maximum beta energy. The fraction of energy lost to radiation is inversely proportional to the charged particle's mass. Bremsstrahlung will therefore be very low for heavy charged particles. Even at 100 MeV, for example, heavy particles dissipate most of their energy through the collisional process.

For mixtures of elements, one determines the "effective atomic number", Z(eff). This can be found by using the following formula, Eqn:

$$Z_{eff} = \frac{\sum f_i Z_i^2}{\sum_i f_i Z_i},$$ where f_i is the fraction by weight of the i the element

Range

A source of monoenergetic alpha particles is directed at a particle detector through an absorber. We know the initially monoenergetic spectrum will demonstrate energy straggling, but the direction of the particles should not be altered and they should make it to detector, provided the thickness of the absorber is less than the thickness required to stop the slowest of the distribution. This is shown as the constant horizontal line in the graph at the bottom of the drawing. Once the absorber thickness is such that some of the slower particles are stopped, the width of the energy straggling curve presents itself in the slope of the fall-off from constant particle number detected down to zero (Figure 5.3).

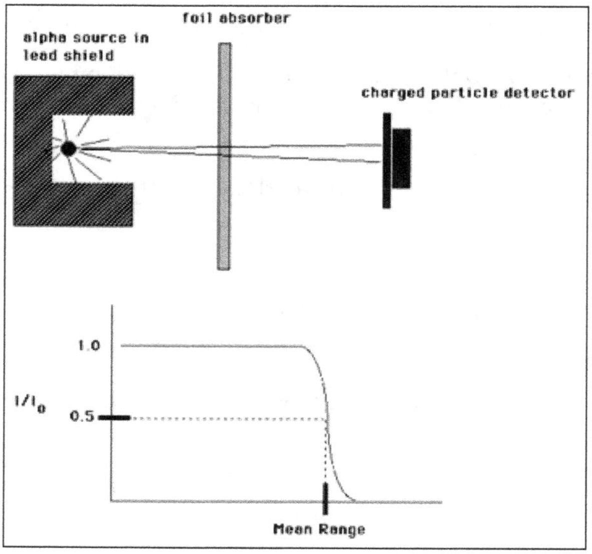

Figure 5.3: Monoenergetic Spectrum.

The mean range is the range at which the number of particles detected is one-half the original value. The same experiment can be performed with a beam of electrons (or a beta emitter). In this case, however, the electrons are more vulnerable to wide-angle scattering even in the thinnest of foils - so wide an angle that the scattered electrons will not be detected. The result is an immediate falloff in number of electrons detected as a function of absorber thickness is as shown in Figure 5.4).

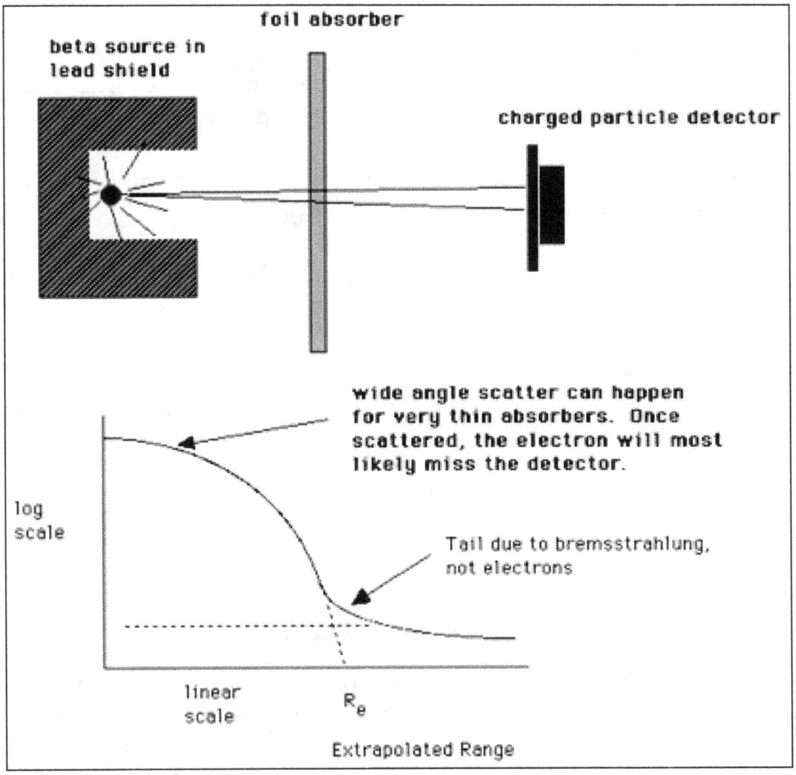

Figure 5.4: Foil Observer.

Factors which Affect Range R

- **Energy:** Range is approximately linear with energy since the Bethe-Bloch equation for stopping power is inversely proportional to E.
- **Mass:** for the same kinetic energy, the electron is much faster than the alpha due to its smaller mass, and therefore the electron has less time to spend near orbital electrons. This reduces the effect of Coulomb interactions (hence stopping power) and increases range.
- **Charge:** the more charge, the more stopping power and the lower range. Range is inversely proportional to the square of the charge of the charged particle.
- For example, a tritium particle with z=1 will have 1/4 the stopping power of a He-3 particle with z=2.
- **Density:** The stopping power increases with increasing density. The range is inversely proportional to the density of the absorbing medium.

Ranges of Heavy Particles

Heavy particles such as protons and alpha particles of certain energy will lose all their energies in a definite distance in a medium, and this distance is called the range. The range is the distance travelled by the particle. The range depends on the energy and mass of the particles. The higher the energy, the larger is the range. When two particles have the same kinetic energy, the heavier one has a shorter range (Figure 5.5).

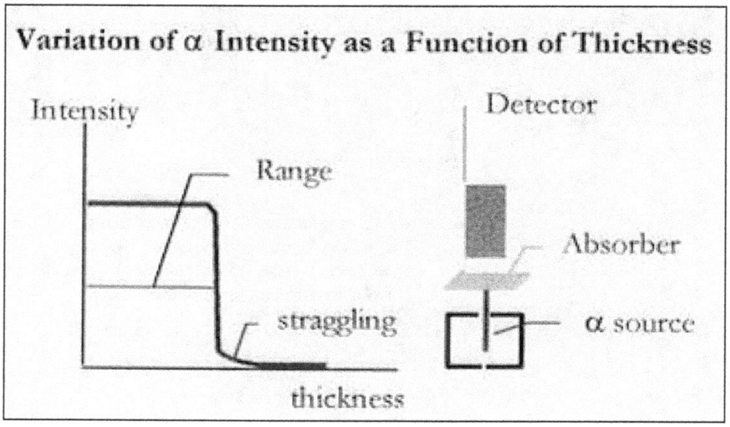

Figure 5.5: Range of Heavy Particles.

A Figure 5.6 shows the graph of relationship among the range and energy such as the one shown here, can be constructed from measurements of some standard sources. If the range of a charged particle source is measured, the energy of the particles can be determined against the chart. The range in aluminium is often used.

The range is a distance or thickness traveled by the particles, but the thickness is often measured in mass per unit area, which is the density times the distance.

5.2 Energy Transfer in Elastic Collision

Simplest atomic collision is the elastic scattering between two particles, under the action of a central force f(**r**) that is a function of inter-particle separation r, but not of angle. Related to the force is the interaction energy U(r).

In elastic scattering the total kinetic energy of the particles before and after the collision is the same, where in an inelastic collision, there is a transfer between kinetic energy and internal energy of the system (electronic, vibrational, rotational energies). Both f(r) and U(r) decay fairly rapidly for r exceeding atomic dimensions, making it possible in most cases to speak of the time before and after the collision, where the interactions are negligibly small (Figure 5.7).

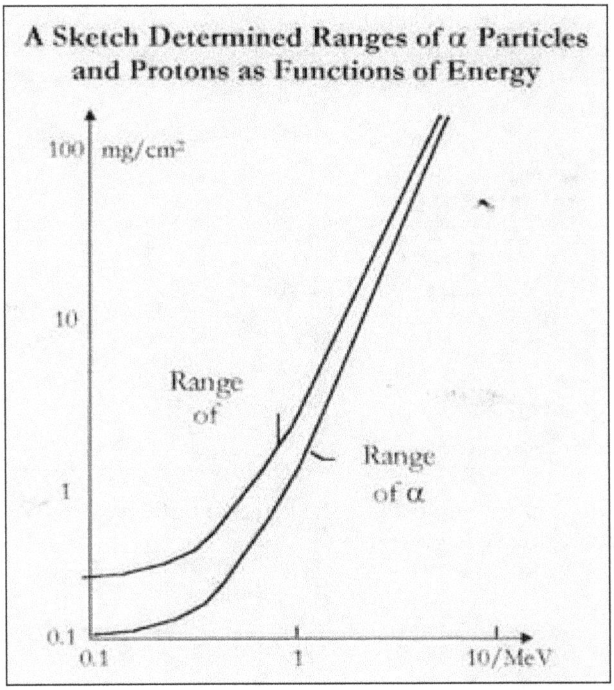

Figure 5.6: Graph of different Range of Particles.

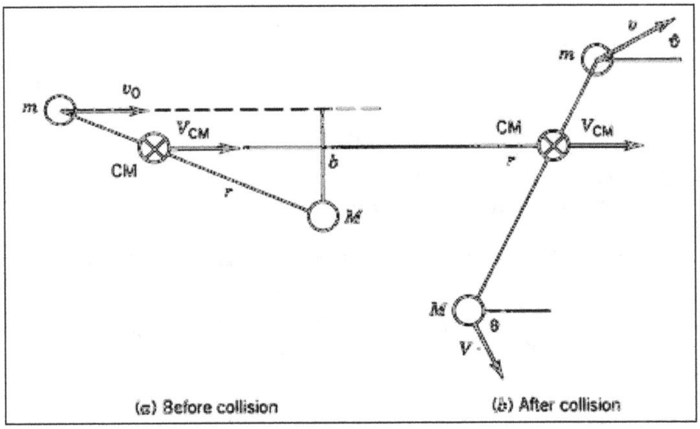

Figure 5.7: Particle Collision.

There are two natural reference frames for describing a collision where one particle moves initially with velocity v_0 and other is stationary, and the center of mass frame, which is at rest before and after the collision and where both particles move. Sometimes one also uses a projectile frame, which moves with the projectile. In this case, the M and m *be* the masses of the projectile and target atom with impact parameter b as closest distance between the projectile and the target in the absence of an interaction. After the collision m and M travel with velocities v, V and lab angles of ϑ, θ, respectively in the lab system, and with angle Q in the CM system.

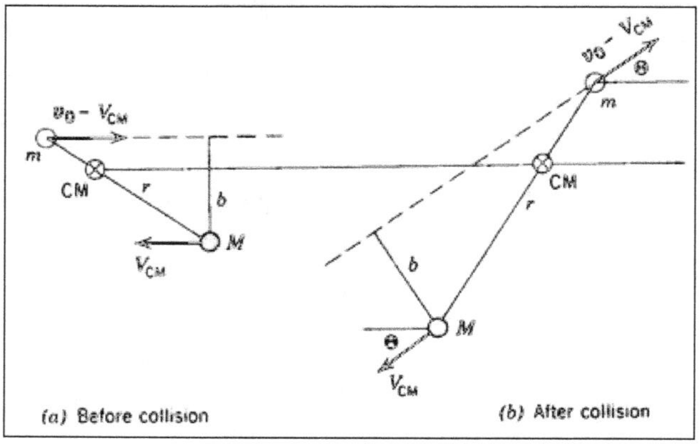

Since the total momentum needs to be the same in both systems, the magnitude of the center of mass velocity is given by

$$(m + M)V_{CM} = mv_0$$

or

$$\text{VCM} = V_0 \frac{M_r}{M}$$

where the reduced mass is defined by:

$$M_r = \frac{mM}{m + M}$$

Note the special cases: M_r is the mass of the lighter particle if there is a large difference in mass. $M_r = 0.5$ m = 0.5 M if m = M.

Since kinetic energy and momentum is conserved in an elastic collision,

$$\frac{1}{2} mv_0^2 = \frac{1}{2} mv^2 + \frac{1}{2} MV^2$$

$$mv_0\ mv \cos \vartheta + MV \cos \theta$$

$$0 = mv \sin \vartheta + MV \sin \theta$$

Note that the inter-atomic forces/potential do not enter in this expression. Solving for the final velocities, one obtains:

$$v2 = v_0^2 \left(1 - \frac{4M_r^2}{mM} \cos^2 \theta\right) = v_0^2 \left(1 - \frac{4mM}{(m+M)^2} \cos^2 \theta\right)$$

$$V = 2v_0 \frac{M_r}{M} \cos \theta = 2v_0 \frac{m}{(m+M)} \cos \theta$$

An important quantity is the elastic energy transfer from the projectile to the target:

$$T = \frac{1}{2} MV^2 = \gamma E_0 \cos^2 \theta$$

$$T = \gamma E_0 \sin^2 \frac{\Theta}{2}$$

Where, E_0 is the initial kinetic energy of the projectile and

$$\gamma = \frac{4mM}{(m+M)^2}$$

Note the special cases:

$$T = 4\frac{m}{M} E_0 \cos^2 \theta \qquad \text{for } m \ll M$$

$$T = 4\frac{M}{m} E_0 \cos^2 \theta \qquad \text{for } M \ll m$$

$$T = 4\frac{1}{2} E_0 \cos^2 \theta \qquad \text{for } m = M$$

The first two conditions say that the energy transfer is very small when the particles differ greatly in mass. The third equation gives the maximum energy transfer, which occurs when the masses are equal. Relationships between lab and CM scattering angles are:

$$\theta = \frac{1}{2} (\pi - \Theta)$$

$$\tan\theta = \frac{\sin \Theta}{1 - \cos \Theta}$$

and

$$\tan\vartheta = \frac{\sin \Theta}{\frac{m}{M} + \cos \Theta}$$

$$v \sin \vartheta = v_0 \frac{M_r}{m} \sin \Theta$$

Now considering special cases:

a) $m \ll M$

$\vartheta \approx \Theta$

That is, when the projectile is much lighter than the target, the lab and CM systems are nearly identical.

b) m = M

$$\vartheta = \frac{\Theta}{2}$$

In this case of identical masses, there is no backscattering in the lab system. Head-on scattering (Q = p) gives J = p/2, q = 0).

c) m > M

There is a maximum scattering angle. As J increases to

$$\vartheta \max = \sin^{-1}(M/m)$$

Q increases from 0 to

$$\Theta = \cos^{-1}(-M/m)$$

d) $m >> M$

$$\vartheta \approx \frac{M}{m} \sin\Theta$$

5.3 Momentum Loss of Heavy Charged Particles

The knowledge of scattering is important in shielding when particles didn't scatter. Many particles of interest in real-world shielding problems do not have absorption event as their first interaction. Many particles bounce around for dozens or hundreds of collisions, resulting in changing direction and losing energy, till they are finally absorbed at an energy far below their original energy. With this, particles penetrate in thick shields, bouncing around the corners, to get instrument penetrations by walls and causing host of other analysis difficulties. As neutron and photon absorption occurs more readily at lower energies, the general strategy in shielding high energy sources is to mix effective slowing-down materials with effective absorption materials.

5.3.1 Momentum

As per equations of conservation of momentum, basic relation involves balancing the three vectors P, P′ and Pr of the particle before the collision and particle after the collision. The laboratory system is used for these vectors.

The three terms which are analogous to momentum terms E, E′, and T for energy of particle before the collision, particle after the collision and target after the collision has another term, Q, representing non-kinetic extra energy that is added to or subtracted from the kinetic energy due to nuclear reactions and excitations.

Reactions that have no Q value are called **elastic** collisions: they correspond to our classic thinking about elasticity since all of the energy is accounted for in the motion of the particles. Reactions with a Q value are called **inelastic** collisions.

Usually the Q value is negative because the target nucleus is left in an excited state, robbing energy from the motion of the particles. In the reactions we are examining in this section, only neutron collisions can be inelastic, since free electrons cannot be excited.

5.3.2 Momentum and Energy

The equation relating energy and momentum is given by equation:

$$E = \sqrt{p^2c^2 + m^2c^4} - mc^2$$

Such equation is applied directly to particles with relativistic energies, but we are more interested in two simpler types of particles: photons and non-relativistic neutrons.

So, for photons, equation simplifies to:

$$E = p^*c$$

Or

$$p = E/c$$

For non-relativistic particles, reducing it to form by substituting

$$p = m^*v$$

to get:

$$E = \sqrt{m^2v^2c^2 + m^2c^4} - mc^2$$

$$= mc^2\sqrt{1 + v^2/c^2} - mc^2$$

and using the first two terms of the Taylor expansion of the square root:

$$\sqrt{1+\varepsilon} = 1 + 1/2\varepsilon - 1/8\varepsilon^2 + \ldots\ldots$$

to arrive at:

$$E = 1/2mv^2$$

5.4 Inelastic Scattering of Light Particles-Radiation Loss

Inelastic scattering is a fundamental scattering process in which the kinetic energy of an incident particle is not conserved. In an inelastic scattering process, some of the energy of the incident particle is lost or increased. Although the term is historically related to the concept of inelastic collision in dynamics, the two concepts are quite distinct; inelastic collision in dynamics refers to processes in which the total macroscopic kinetic energy is not conserved. In general, scattering due to inelastic collisions will be inelastic, but, since elastic collisions often transfer kinetic energy between particles, scattering due to elastic collisions can also be inelastic, as in Compton scattering.

When at least one of the colliding particles is composite - composed of smaller constituents - relative motion of the constituents can take up some of the initial energy. This is one of the processes that can take place when, for example, a high-energy particle hits an atom and kicks one of its orbital electrons out of its place near the atomic nucleus. This phenomenon, called ionization, is used in a wide range of particle detectors, from hand-held Geiger counters to the large drift chambers used in particle physics experiments (Figure 5.8).

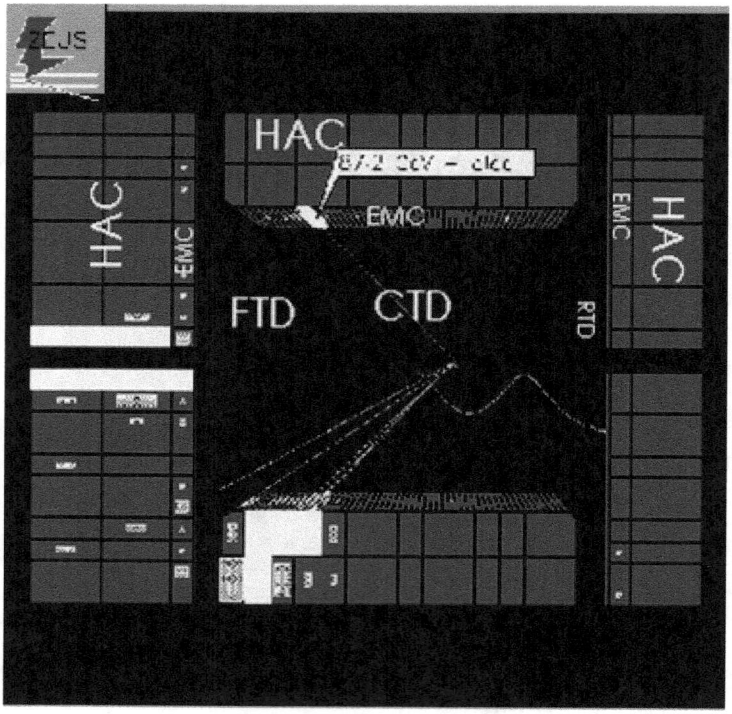

Figure 5.8: Detector.

As in the case of macroscopic bodies, a composite particle may break up into its constituents. In this case, the initial energy is shared among the kinetic energies of all the final particles, and some of it also goes to sever the bonds that initially held the constituents together.

When the constituent particle is a hadron - generally, a particle made of quarks - its breakup results in an additional process that is unique to the quantum mechanical regime of particle physics and does not take place in collisions of macroscopic bodies: the creation of more quarks and anti-quarks. This happens because the attractive force that binds the quarks together increases roughly in proportion to the distance between them, like a spring. As the quarks fly apart and the distance between them increases to about the size of the proton (10^{-13}cm), the energy stored in that spring is enough to create the mass of a quark-anti-quark pair. The energy cost to create this pair is, of course, $E=2mc^2$, where m is the mass of each of the particles in the pair.

When the electron is more energetic, more quark-anti-quark pairs can be created, resulting in the production of more hadrons. The following picture shows such an event, recorded by the ZEUS detector at the HERA electron-proton storage ring. The electron beam is coming in from the left of the picture and the proton beam is coming in from the right. The electron is scattered to the top left part of the picture, leaving a track in the drift chamber. The small yellow blob at the end of the track indicates that the electron deposited all its energy in the electromagnetic calorimeter. The particles created in the breakup of the proton go mostly to the bottom left, where we can see a jet containing several charged particles and energy deposited in the hadronic calorimeter. Note that the calorimeter also records the energy deposited by neutral particles, which would not be detected by the drift chamber.

During the collision, the electron and positron change direction very rapidly, undergoing high acceleration. When a charged particle is accelerated, it emits photons - particles of electromagnetic radiation, such as visible light. In the high energy range we work with in particle physics, such photons are called gamma-rays. Particle collisions in which a photon is emitted (radiated) are called radiative collisions.

The photons emitted in the radiative e-e+ collision usually have energies that are low compared to those of the incoming particles. Very often, they also have a very small angle with respect to one of the incoming particles. In this case, they go down into the beam type and are not detected by the detector. In other cases, these photons have a small angle with respect to one of the outgoing charged particles, with a relatively low energy. As a result, the photon and the charged particle deposit their energies in the same cluster of calorimeter crystals, making it hard or impossible to detect the photon by identifying two separate signatures in the calorimeter (Figure 5.9).

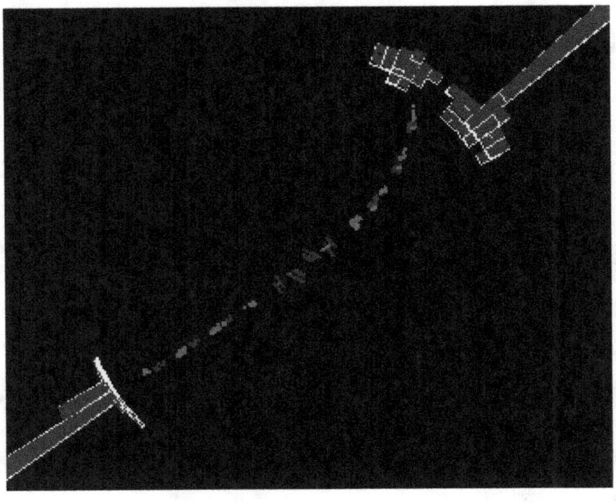

Figure 5.9: Calorimeter.

But often enough, the radiated photon is so energetic that the outgoing charged particle is left with relatively little kinetic energy, and therefore curves in the magnetic field before reaching the calorimeter. The photon is not charged and therefore does not curve, resulting in enough separation between the charged particle and the photon that they can be separately identified in the calorimeter.

5.5 Attenuation Coefficients

Attenuation coefficient of the volume of a material characterizes how easily it can be penetrated by a beam of light, sound, particles, or other energy or matter. A large attenuation coefficient means that the beam is quickly attenuated as it passes through the medium, and a small attenuation coefficient means that the medium is relatively transparent to the beam. The SI unit of attenuation coefficient is the reciprocal metre (m^{-1}).

5.5.1 Attenuated Intensity

When a narrow beam of mono-energetic photons, such as X-ray, γ rays, penetrates a surface ofa material; it experiences a change in its intensity, owing to absorption or scattering that occurduring the travel of the beam across the material (Figure 5.10).

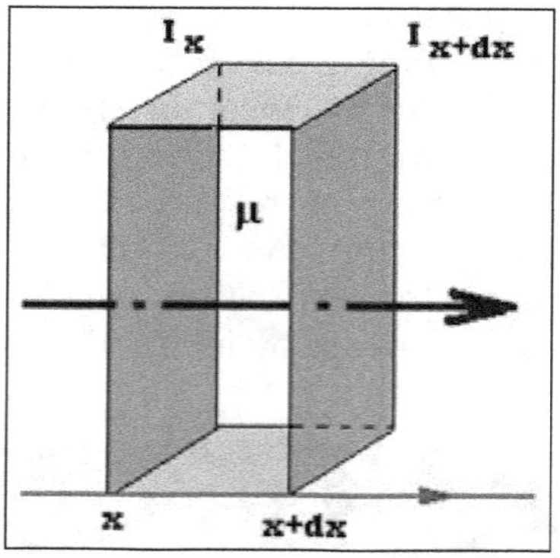

Figure 5.10: Material Surface.

The expression of the intensity of the beam along an axis parallel to the direction that photons beam can be calculated as:

$$dI_x = I_{x+dx} - I_x = -(I_x - I_{x+dx}) = -\text{ coefficient } dx\, I_x$$

Or:

$$dI_x/I_x = -\text{ coefficient } I_x\, dx.$$

Integrated with respect to the thickness of material traversed "dx", and letting coefficient = μgives:

$$\ln I(x) = -\mu x + \text{constant} \rightarrow I(x) = \text{Constant } e^{-\mu x}$$

At x = 0, I(o) = I_o; then, at the traveled distance "x", we have

$$I(x) = I_o e^{-\mu x}$$

5.5.2 Linear Attenuation Coefficient

The coefficient μ must have the dimension of length^{-1}. It is called the linear attenuation coefficient. It is generally expressed in cm^{-1}. The linear attenuation coefficient μ is defined as the fraction of an incident beam of photons that is absorbed or scattered per unit thickness of the target absorber.

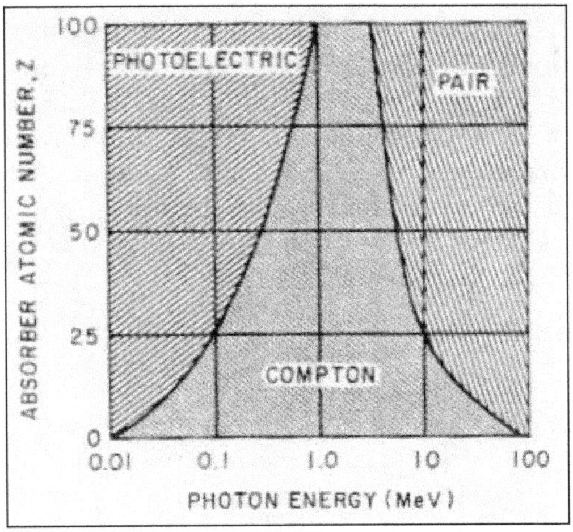

The photoelectric effect dominates at low photon energies (< 26 keV) in soft tissue. With higher energy photons at low Z materials, Compton scattering dominates. Rayleigh scattering comprises about 10% of the interactions in mammography and 5% in chest radiography.

The average distance traveled by a photon between two successive interactions is called Mean Free Path (MFP), and can be calculated from linear attenuation coefficient: $\lambda = 1/\mu$.

5.5.3 Mass Attenuation Coefficient

The coefficient μ is used as μ/ϱ and called mass attenuation coefficient, where ϱ is the density of the target material. It is usually expressed in units of cm^2/g. It is obtained by measuring the values of Io, I and the thickness "x".

5.5.4 Total Mass Attenuation Coefficient

The mass attenuation coefficient is due to absorption coefficient related to the incoherent processes, as the phoelectric effect, Compton Effect and pair production; and scattering coefficient due to coherent effect as Rayleigh scattering processes.

No we write:

μ/ϱ (Total) = μ/ϱ(absorption) + μ/ϱ(scattering), that is:

μ/ϱ(Total)=μ/ϱ(Photoelectric)+μ/ϱ(Pair-Production)+μ/ϱ(coherent:Rayleigh) + μ/ϱ(incoherent:Compton)

For mixtures and compounds, the values of the mass attenuation coefficients are obtained by additivity: $\mu/\varrho = \sum wi\ (\mu/\varrho)i$, where wi is the fraction by weight of the ith atomic constituent.

5.6 Classical (Thomson) Scattering from Single Electron

One of the most powerful methods of diagnosis is to use the scattering of electromagnetic radiation from the plasma. The attractiveness of this diagnostic derives from two main features. First, it is, for all practical purposes, no perturbing method, requiring only access of radiation to the plasma. Second, it offers the potential of determining detailed information about the distribution function of electrons and sometimes even of the ions too. These advantages are sufficient to offset the fact that the measurements are generally very difficult technically to perform. Electromagnetic wave scattering diagnostics are now widespread, especially in hot plasma experiments.

The elastic scattering of X-rays from free electrons is known as Thompson scattering.

Consider a beam of unpolarised X-rays with electric field vector E_o travelling along the z direction which is incident on a free electron positioned at the origin as shown in figure. The electric field vector in just one direction E_o, can be divided into two components E_{ox} and E_{oy}. The component E_{ox} produces a force which acts on the electron and makes it oscillate. In the x direction this force will be eE_{ox} where e is the charge of an electron.

If the mass of the electron is m, then the equation of motion may be written

$$x = eE_{ox}/m * \exp(\omega t)$$

The accelerated electron in turn radiates electromagnetic radiation. At a large distance R from the electron and at an angle Φ to the z direction the amplitude of the electric field vector of the scattered radiation in the xz plane is given by

$$E_x = -e/4\pi\varepsilon c^2 R * x(t-R/c)\cos\Phi$$

where

ε = permittivity of free space

c = speed of light

(t-R/c) = retarded time

In the Figure 5.11, electric field vector at O in one particular direction E_o is considered as having two components in the x and y directions E_{ox} and E_{oy}. The corresponding electric field vectors of the scattered radiation in xz plane at a distance R from O at an angle Φ to z are Ex and Ey respectively. From eqs, we can write the amplitude of scattered radiation due to the x component as

$$E_x = -e^2 Eox/4\pi\varepsilon mc^2 R * \cos\Phi$$

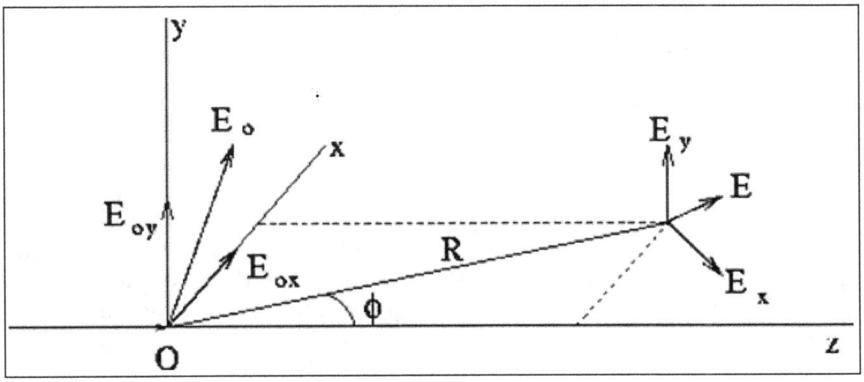

Figure 5.11: Incident Unpolarised Beam of X-rays Travelling Along *z* is Incident on Free Electron at Origin.

The equivalent expression for the perpendicular component Ey has no Φ dependence and we may therefore write the electric field in one direction perpendicular to the direction of motion of the scattered beam as

$$E^2 = e^4/(4\pi\varepsilon)^2m^2c^4R^{2*}\ E^2oy + E^2ox\cos\Phi$$

The total electric field at the point of observation may be found by averaging over all directions of Eo in the xy plane to give

$$\langle E^2\rangle = e^4/(4\pi\varepsilon)^2m^2c^4R^2 \langle E^2o\rangle (1 + \cos^2\Phi/2)$$

The term $(1 + \cos^2\Phi/2)$is known as the polarisation correction for an unpolarised incident X-ray beam.

In terms of X-ray beam intensities, which are proportional to the square of the amplitudes of the electric field vector of the radiation, we can therefore write

$$I = Io^*\ e^4/(4\pi\varepsilon)^2m^2c^4R^2 (1 + \cos^2\Phi/2)$$

For an atom containing Z free electrons the scattering in the forward direction (scattering angle $\phi = 0$) is Z times the intensity given by above equation as scattered radiation from each electron in the atom will be in phase at $\phi = 0$. However, as ϕ increases from zero the scattered X-rays from the electrons begin to interfere destructively and the strength of the overall scattering falls off with the increasingϕ.

The form of this fall-off is dependent on the charge distribution inside the atom and is represented by an atomic form factor, or scattering factor, *f* defined as the ratio between the amplitude scattered by an atom and that scattered by a free classical electron when all other conditions remain unchanged.

If the incident X-ray energy is much greater than the binding energies of electronic states within the atom then the electrons may be considered to be free and Thompson scattering holds. Assuming that the *n*th electron in an atom has a spherically symmetric charge distribution pn(τ) then the atomic scattering factor for the Thompson scattering case, *fo*, is given by

$$fo = \sum \int \infty\, 4\pi r2\ \varrho n(r)\ \sin sr/sr{*}dr$$

Where

$$s \ = 4\pi \sin \theta_B/\lambda$$

For the majority of cases the two assumptions that the electrons are free and that their charge distributions are spherical have been found to be adequate for the purposes of crystallographic investigation on structures containing atoms of low atomic number.

The largest deviations from the Thompson scattering approximation occur when the incident X-ray energy is close to the binding energies of electrons within the atoms of the structure, *i.e.*, at energies around the X-ray absorption edges of the atom. The X-ray wavelengths used for crystallographic structure determination are typically of similar size to the dimensions of the atoms being investigated (of the order of 1 Å). The corresponding X-ray energies coincide with the absorption edges of heavier elements. Thus for these elements the Thompson scattering approximation is no longer valid. This change in behaviour away from Thompson scattering at energies around absorption edges is known as anomalous scattering.

5.7 Incoherent/Compton Scattering

In quantum mechanics, scattering of X-rays by free electron is given by Klein-Nishina formula. If we assume unpolarized X-rays and unaligned electrons, this formula can be approximated as follows for X-ray energies below 100 keV:

$$d\sigma_{KN}/d\Omega \cong \frac{r_e^2(1+\cos^2\theta)}{2[1+K(1-\cos\theta)]^2},$$

where

$K = E/mc^2$ the photon energy measured in units of electron rest energy.

The total cross section is approximately

$$\sigma_{KN} \cong 8\pi r_e^2 \frac{(1+2k+1.2\,k^2)}{3(1+2k)^2}$$

Here, for very low energies ($k \rightarrow 0$), we recover Thomson cross section. The real difference comes when we deal with atoms. In that case, if the scattering leaves the atom in the ground state, we deal with coherent scattering, whereas if the electron is ejected from the atoms, the scattering is (incoherent) Compton scattering.

At high energies, the total Compton cross section approaches $Z\sigma_{KN}$. At low energies and small scattering angles, however, binding effects are very important, the Compton cross section is significantly reduced, and coherent scattering dominates as shown in figures below.

The scattered X-ray suffers an energy loss, which (ignoring binding effects) is given by

$$E/E = 1/[1 + k(1 - \cos\theta)]$$

or, in terms of the wavelength shift,

$$\lambda^1 - \lambda = \lambda_c (1 - \cos\theta),$$

where

$$\lambda_C = h/mc = 2.46 \times 10^{-12} \text{ meter}$$

The kinetic energy of the recoil electron is just the energy lost by the photon in this approximation:

$$E_e = \frac{k(1 - \cos\theta)}{1 + K(1 - \cos\theta)},$$

In the Figure 5.12, total photon cross section σ_{tot} in carbon as a function of energy shows contributions of different processes: t, atomic photo-effect (electron ejection, photon absorption); σ_{coh}, coherent scattering (Rayleigh scattering—atom neither ionized nor excited); σ_{incoh}, incoherent scattering (Comp- ton scattering off an electron); k_n, pair production, nuclear field; k_e, pair production, electron field; σ_{ph}, photonuclear absorption.

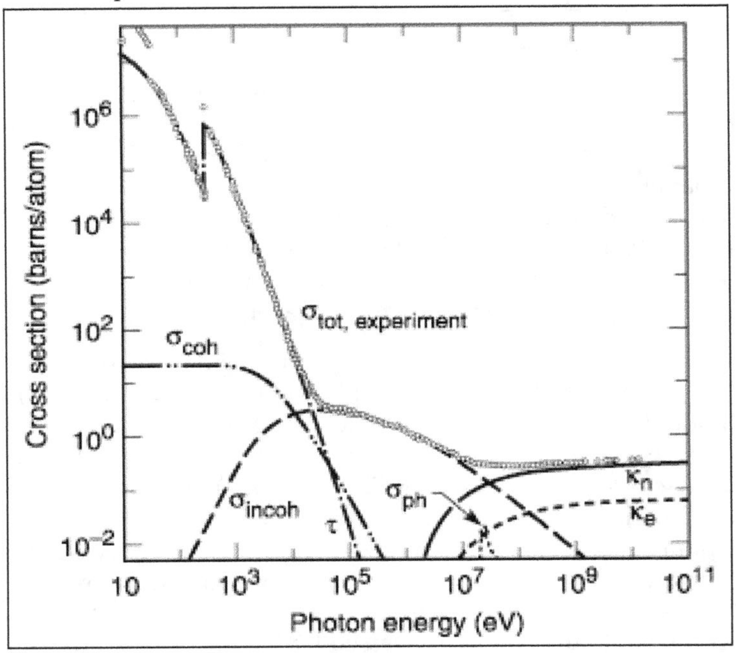

Figure 5.12: Total Photon Cross Section σ_{tot} in Carbon as a Function of Energy.

5.8 Photoelectric Absorption

Photoelectric absorption is the process which produces the high-speed ionizing particle which causes the scintillation to appear in the phosphors of scintillation detectors. It produces an effect in the formation which influences some density logging measurements (Figure 5.13).

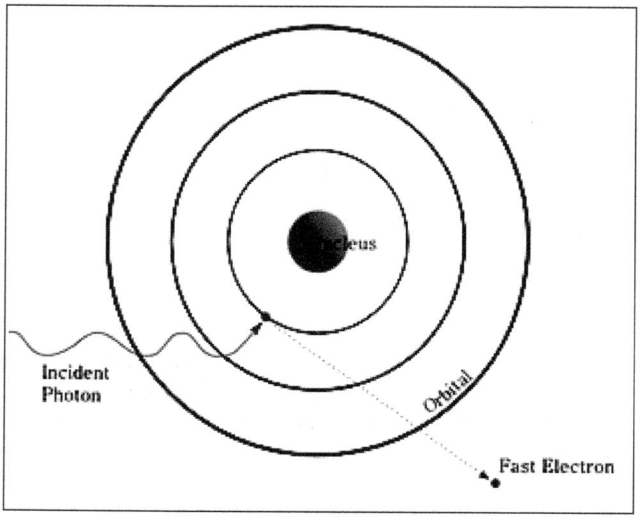

Figure 5.13: Photoelectric Absorption.

An atom completely absorbs a photon, which then disappears; this excess energy provided to the atom ultimately results in ejection of an orbital electron. The ejected electron is known as a photo electron. Electrons have binding energies of orbit, with outer shells having less energy than those closer to the positive nucleus. When the initial orbital vacated is not that of an outer valance electron, the atom remains in a high-energy state until an outer orbital electron shifts to fill its incomplete inner shell. This shift is accompanied by emission of a characteristic X-ray. Photoelectric Absorption is an important interaction for low energy photons (<100 KeV).

In photoelectric absorption, a photon disappears being absorbed by an atomic electron. The process results in ionization by subsequent ejection of the electron from the atom. The energy of the liberated electron is the difference between the photon energy and the energy needed to extract the electron from the atom *i.e.* the binding energy of the electron. The recoil momentum is absorbed by the nucleus to which the ejected electron was bound. If the resulting photoelectron has sufficient amount of kinetic energy, it may be a source of a secondary ionization occurring along its trajectory, and in the case of the semiconductor material, it may create further e-h pairs. If the electron does not leave the detector the deposited energy corresponds to the energy possessed by the incident photon. This feature of the photoelectric effect allows calibrating the gain of the detector chained with its readout system if the energy required to create a single e-h pair is known.

The range R of the electron having the kinetic energy E is of the order of some micrometers, as given by the follow equation:

$$R\,[\mathrm{um}] = 40.8\ 10^{\wedge(-3)} \times (\ \mathrm{E[keV]}\ )^{\wedge 1.5}$$

Thus, the cloud of generated charge is confined close to the photon absorption point. The clear image may be smeared by escape photons, which can leave the detector volume leading to less amount of energy deposited. These photons are actually the fluorescence photons emitted by de-exciting atoms. Photons of fluorescence radiation are emitted by atoms after the ejection of a deep shell (K, L) electron. In this incident, photon creates a vacancy in the shell, thus leaving an atom in an excited state. Then, the vacancy can be filled by an outer orbital electron, giving rise to the emission of the characteristic X-rays photons of the fluorescence radiation. The missing energy, which is conveyed by the escape photons leads to, so called escape peaks in the measured energy spectrum. Photon interaction coefficient for photoelectric absorption depends strongly on the atomic number of the absorbing material. The relevant cross section increases roughly as $Z^{\wedge 3}$. For silicon, the photoelectric effect is a dominant process for photon energies below 100 keV.

The Compton scattering, instead of photoelectric effect, involves the free electrons. In matter of course, the electrons are bound to an atom; however, if the photon energy is high with respect to the binding energy, this latter energy can be ignored and the electrons can be treated as essentially free. When Compton scattering occurs, the electron is scattered away in conjunction with a new photon that have a lower energy than the incoming one.

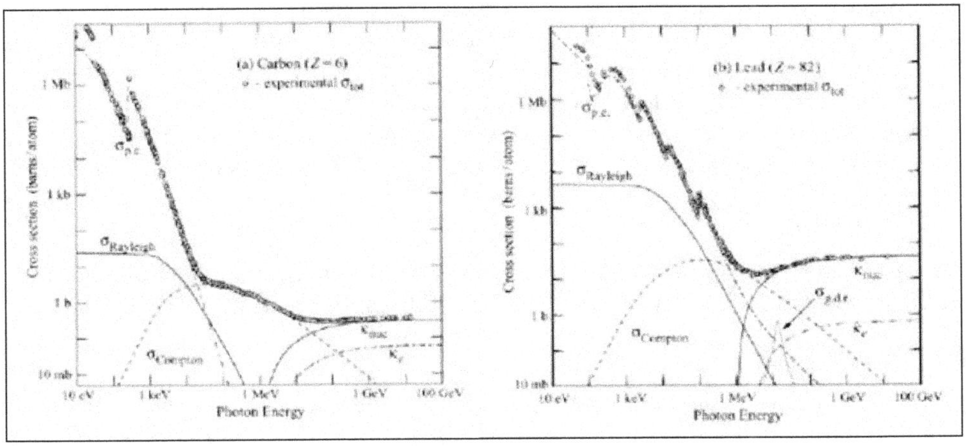

Figure 5.14: Cross Sections of Photons in Carbon and Lead.

In Rayleigh scattering the photon interact with the whole atom and the only effect of this interaction is a deflection of the incoming photon; it does not participate to the absorption and for most purposes can be neglected.

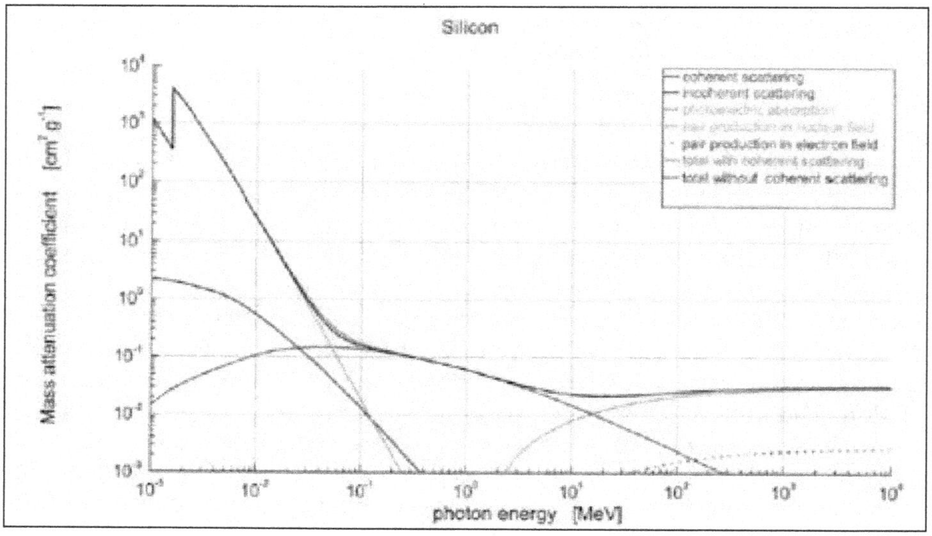

Figure 5.15: Mass Attenuation Coefficient of the Silicon and its Components.

At very high energy, another effect starts to be relevant: the pair production. In this process the photon interacts with an electron or a nucleus producing a positron-electron pair. In order to produce the pair, the photon must have at least an energy of 1.022 MeV. In above figure, with k_{nuc} *and* $k_{e'}$ are shown the two components of the pair production cross section, respectively for the interaction with nuclei or electrons. Another possible interaction, but usually negligible compared to the previous ones is the Photonuclear reaction, in this case the photon interact directly with the nucleus. The related cross section is shown in Figure 5.14 in dotted line ($\sigma_{g.d.r.}$). The above cross section in barns/atom (1barn = 10^{-24} cm^2, approximately the section of an uranium nucleus) expresses the probability of an interaction. A more suitable quantity, often used to characterize the absorption of a photon shower, is the mass attenuation coefficient. The mass attenuation coefficient is defined as:

$$\mu m = \eta \mathrm{A}\ \sigma \mathrm{tot}/\varrho$$

Where, ϱ is the density of the material. The Figure 5.15 shows the mass attenuation coefficient of the silicon with the indication of its different components.

References

J. H. Hubbell, W. J. Veigele, E. A. Briggs, R. T. Brown, D. T. Cromer, and R. J. Howerton, "Atomic Form Factors, Incoherent Scattering Functions, and Photon Scattering Cross Sections," J. Phys. Chem. Ref. Data 4, 471 (1975).

J. H. Hubbell, H. A. Gimm, I. , "Pair, Triplet, and Total Atomic Cross Sections (and Mass Attenuation Coefficients) for 1 MeV–100 GeV Photons in Elements Z = 1 to 100," J. Phys. Chem. Ref. Data 9, 1023 (1980).

R. D. Evans, The Atomic Nucleus (Kreiger, Malabar, FL, 1982); R. D. Evans, "The Compton Effect," in S. Flugge, Ed., Handbuch der Physik, vol. 34 (Springer-Verlag, Berlin, 1958), p. 218; W. J. Veigele, P. T. Tracy, and E. M. Henry, "Compton Effect and Electron Binding," Am. J. Phys. 34, 1116 (1966).

6

Basic of Detectors

6.1 History and Future Perspectives

The interactions of alpha, beta, and gamma radiations with matter produce positively charged ions and electrons. Radiation detectors are devices that measure this ionization and produce an observable output. Early detectors used photographic plates to detect "tracks" left by nuclear interactions. The cloud chambers, used to discover sub-nuclear particles, needed photographic recording and a tedious measurement of tracks from the photographs.

Advances in electronics, particularly the invention of the transistor, allowed the development of electronic detectors. Scintillator-type detectors use vacuum tubes to perform the initial conversion of light to electrical pulses. The amplification and storing these data follow the advances in transistor electronics. Miniaturization in electronics has revitalized types of gas-filled detectors. These detectors were developed as "single element" detectors and now have been revived into "multiple element" detectors with more than one thousand elements. Advances in materials, particularly ultra-pure materials, and methods of fabrication have been critical to the creation of new and better detectors.

Particle detectors, also called radiation detectors, are instruments designed for the detection and measurement of subatomic particles such as those emitted by radioactive materials, produced by particle accelerators or observed in cosmic rays. Such particles include electrons, protons, neutrons, alpha particles, gamma-rays, and numerous mesons and baryons. Most detectors utilize in some way the ionization produced when these particles interact with matter.

6.2 Radiation Detectors

Radiation is energy traveling in the form of particles or waves in bundles of energy called photons. Some everyday examples are microwaves used to cook food, radio waves for radio and television, light, and X-rays used in medicine.

Radioactivity is a natural and spontaneous process by which the unstable atoms of an element emit or radiate excess energy in the form of particles or waves. These emissions are collectively called ionizing radiations. Depending on how the nucleus loses this excess energy either a lower energy atom of the same form will result, or a completely different nucleus and atom can be formed.

Ionization is a particular characteristic of the radiation produced when radioactive elements decay. These radiations are of such high energy that when they interact with materials, they can remove electrons from the atoms in the material. This effect is the reason why ionizing radiation is hazardous to health, and provides the means by which radiation can be detected (Table 6.1).

Table 6.1: A Partial List of Detectors Used in Nuclear Science. Some detectors can be used only in a limited energy range.

Particle Type	Detector Type	Features
	Charged	
Protons, nuclei, electrons, or pions	Geiger-Müller counters	Portable radioactivity detector
	Gas ionization counters	Gas-filled chamber in an electric field
	Multiwire chambers	Good position resolution
	Semiconductor detectors	Good energy resolution
	Magnetic spectrometers	Good momentum resolution
	Scintillators and photomultipliers	Good timing resolution
	Cerenkov detectors	Good particle identification
	Neutral	
Photons	Scintillators and photomultipliers	Good timing, moderate energy resolution
	Germanium semiconductor crystals	Good energy resolution
Neutrons	Liquid scintillator or BF3tubes	Via fission, capture gamma-rays, or proton collisions
Neutrinos	Cerenkov detectors nuclear reactions	Via neutrino-electron interactions detect resultant radiation

Features

In a wide category of detectors, when a particle or a radiation quantum interacts with material according to one of the previous discussed mechanisms, the net result is the appearance of an amount of electric charge inside the detector active volume. Applying an electrical field within the detector, a flow of charges is originated. The time required for the charge collection is related to the particular detector, in fact, it depends on the charge carrier mobility and on the average distance that must be travelled before arrival at the collecting electrodes. Therefore, the response of a detector to a single particle or to an interacting quantum of radiation is a current that flows for a time equal to the charge collection time. Obviously, the time integral over the current duration is the charge generated during the interaction process. In this analysis a low interaction rate is considered, so that each individual interaction gives rise to a current that is distinguishable from the others. In real situation, current flows in the detector from more than one interaction at a given time.

Detector Working Modes

Three different detector working modes can be considered: pulse mode, current mode and mean square voltage mode. If the response time of the measuring device connected across the detector output terminals is long compared with the average time spacing between current pulses, then a current is recorded that is given by the mean rate of charge formation averaged over many individual radiation quanta. This mode of operation is called current mode. The average current represents the product of the average event rate and the charge created per event. However, different currents will result from radiations that have equal interaction rates but deposit a different average energy per interaction.

In mixed radiation environments, when the charge amount produced by various radiation types is different, detectors operating in mean square voltage mode are useful. In such devices, the mean square of the time variance of the detector current is recorded which is proportional to the event rate and the square of the charge produced in each event.

In the pulse mode operation, each quantum of radiation that interacts with the detector has to be measured. The output of such type of detector is a sequence of individual signal pulses, each representing the result of a single quantum of radiation within the detector. The rate at which such pulses occur, give a measure of the radiation interactions inside the detector while the amplitude of each pulse is related to the amount of charge generated during each interaction.

Pulse mode operation is the most common choice for different radiation detector applications; in fact it presents several advantages (for instance a higher sensitivity) in comparison with the other two working modes.

Pulse Height Spectra and Energy Resolution

In the pulse mode detector, the amplitude of the various pulses is different because of differences in radiation energy or fluctuations in detector response to a monoenergetic radiation. The pulse amplitude distribution gives information about the incident radiation. In Figure 6.1 shown with example of differential pulse height distribution is indicated: the value of dN/dH is plotted vs. the pulse amplitude where dN is the differential number of pulses with an amplitude within the differential increment dH. The number of pulses having an amplitude between H_a e H_b is obtained adopting the following expression:

$$\int_{H_b}^{H_a} \frac{dN}{dH} dH$$

In many applications, the radiation detector task is the energy distribution measure of the incident radiation. The capability of detectors to distinguish between two particles or photons with different but close energies is indicated as the energy resolution and denoted with R.

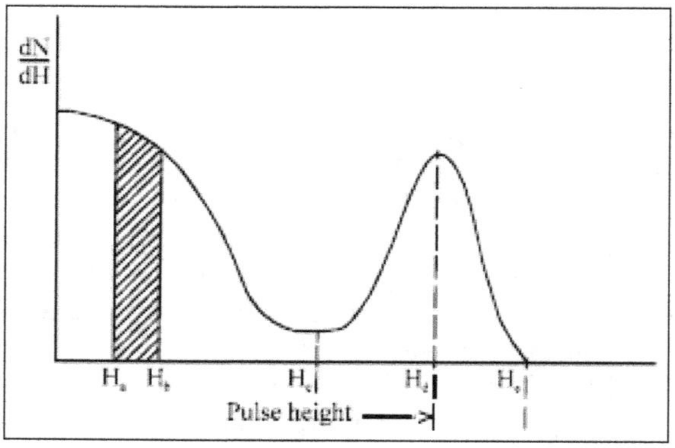

Figure 6.1: Differential Pulse Height Distribution Plot.

A large amount of fluctuations in the number of produced charges is recorded even if radiation beams with equal energy are adopted. There is a high number of potential sources of fluctuations in the response of a detector that result in imperfect energy resolution.

They can be produced by drift of the detector operating characteristics, random noise sources of the instrumentation system and statistical noise derived from the fact that the charge generated within the detector by a radiation quantum is not a continuous variable but represents a discrete number of charge carriers.

The energy resolution is defined as the Full Width at Half Maximum (FWHM) divided by the location of the peak centroid H_0, having indicated with FWHM the width of the distribution at a level that is half the maximum ordinate of the peak shown in Figure 6.2. It is evident that the smaller the value of R, the better the detector will able to distinguish between radiations having energies lie near each other. Assuming that the formation of each charge carrier is a poisson process, it was demonstrated that for detector having a linear response, R results:

$$R = \frac{2.35}{\sqrt{N}}$$

where, N is the average number of generated charge carriers.

For some type of detectors, experimental results show that the obtained R-values are lower than the values predicted by above equation. This difference indicates that the processes which give rise to the generation of charge carriers are not independent and therefore the total charge produced cannot described by a simple poisson process. A corrective parameter, called the Fano factor and indicated with F, is introduced to quantify the difference of the observed values from the number of charge carriers obtained considering a pure Poisson statistic. Introducing the Fano factor, expression above will become:

$$R = 2.35\sqrt{\frac{F}{N}}$$

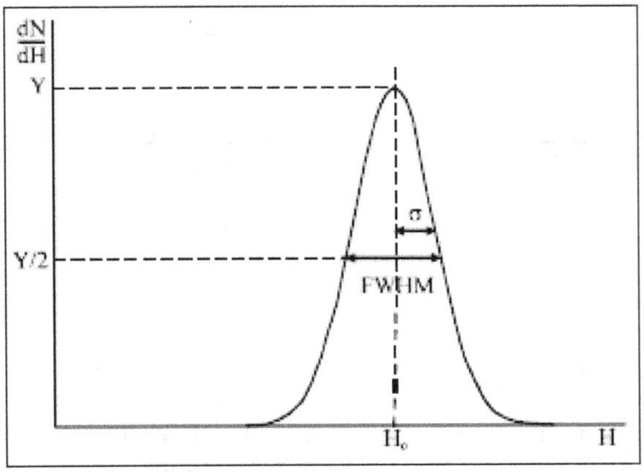

Figure 6.2: Parameter Identification for the Energy Resolution Definition.

Detection efficiency: Charged radiation interacts with detector producing ionization or excitation of material nuclei as it enters in the active volume. Therefore, after a typical particle is travelled a small fraction of its range, a lot of pairs is created to ensure the presence of a signal pulse large enough to be detected. In this situation no charged particles reaching the detector active volume are lost. The detector has a 100% detection efficiency.

On the contrary, uncharged radiations can travel some distance before interacting with a material atom; therefore detector have, generally, an efficiency value less than 100%.

In order to relate the number of counted pulses to the number of hitting photons, the absolute (ε_{abs}) and the intrinsic (ε_{int}) efficiency are evaluated with the following expression:

$$\varepsilon_{abs} = \frac{\text{Number of pulses recorded}}{\text{Number of quanta generated by source}}$$

$$\varepsilon_{int} = \frac{\text{Number of pulses recorded}}{\text{Number of quanta hitting the detector}}$$

The intrinsic efficiency, which is the most used parameter, depends on the detector material, the radiation energy, the detector thickness in the direction of incident radiation. Moreover, a slight dependence on the distance between the source and the absorber material is observed as the average path length of the ray inside the detector changes somewhat with this spacing.

Dead time: The dead time is the time during which a detector is not able to detect a next coming particle, in fact two events must have a minimum amount of separation in time in order to be recorded as two different pulses. This minimum recording time can be determined by physical processes in the detector or by associated electronics.

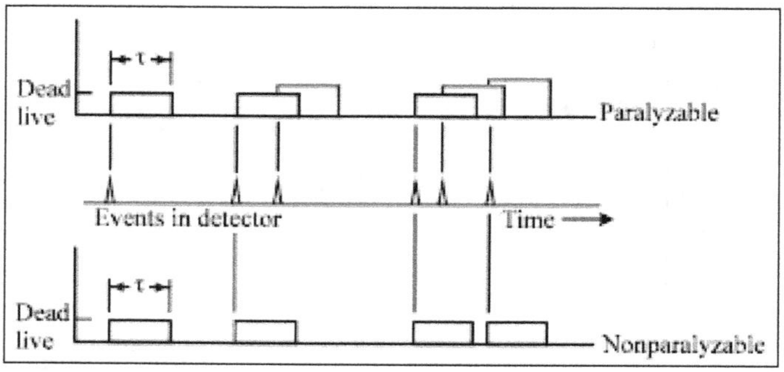

Figure 6.3: Fundamental Assumptions of the Two different Models for Detector Dead Time Behaviour.

For example, when a radioactive sample is counted, the possibility can arise that two interactions in the detector will occur too close together in time to be registered as separate events. Two different models have been proposed to approximate the dead time behaviour of counting systems: paralyzable and non-paralyzable response.

A paralyzable detector is unable to provide a second response until a certain dead time τ has passed without another event occurring. Another event during τ causes the insensitive period to be restarted.

A non-paralyzable detector, on the other hand, simply ignores other events if they occur during τ.

Differences in the two models are illustrated in above figure where central line shows six events as they occur along the horizontal time axis, and the other two axes show the responses of the two types of detectors. Of the six events in this example, the paralyzable counter registers three and the non-paralyzable, four.

Denoting with r_i and r_c the true interaction rate and the recorded count rate respectively, the fraction of time that the non-paraliyzable detector is dead is $r_c\tau$. Therefore, the fraction of time that it is sensitive is $1\text{-}r_c\tau$, which is also the fraction of the number of true events that can be recorded as it is expressed in the following formula:

$$\frac{r_c}{r_i} = 1 - \tau_c\tau(\text{non-paralizable model})$$

In the paralyzable case, dead periods are not always of fixed length. The average number of events that takes place in a time t is $r_i t$. If an event occurs at time t =

0, then the probability that no events occur in time t immediately following that event is given by the Poisson term, with P_0 expressed as:

$$P_0 = e^{-} r_i t$$

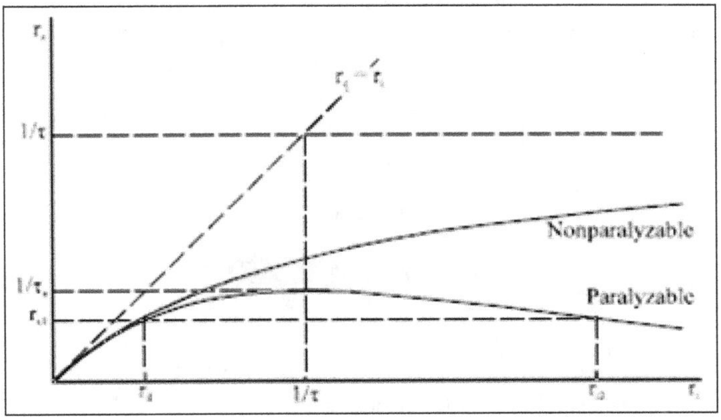

Figure 6.4: Plot of the Observed Rate vs. the True Rate for Non-paralyzable and Paralyzable Models.

Therefore, given an event at time t = 0, the probability that the next event will occur between t and t + dt is:

$$Pt(dt) = rie^{-r_i t}\, dt$$

The probability that a time interval larger than τ will elapse is:

$$\int_{\tau}^{\infty} r_i e^{-r_i t} dt = e^{-r_i \tau}$$

The observed count rate r_c is:

$$rc = r_i e^{r_i \tau} \text{ (paralyzable model)}$$

For low event rates or short dead time ($r_i\tau < 1$), the two models lead the same results:

$$r_c = \frac{r_i}{1 + r_i\tau} \approx r_i(1 - r_i\tau) \text{ (non-paralizable model)}$$

$$rc = r_i e^{-r_i\tau} \approx r_i(1 - r_i\tau) \text{ (paralizable model)}$$

Plotting r_c vs. r_i it is evident that for non-paralyzable model r_c cannot exceed τ^1, which is an asymptotic value, while for paralyzable model r_c reaches a maximum value equal to $e\tau^{-1}$. Increasing the event rate, the measured count rate with a paralyzable system will decrease beyond this maximum and will approach zero, because of the decreasing opportunity to recover between events. With a paralyzable system, there are generally two possible event rates that correspond to a given count rate.

6.3 Types of Radiation Detectors

6.2.1 Scintillation Detectors

Scintillators are solids that give off light when radiation interacts with them. The light is converted to electrical pulses that are processed by electronics and computers (Figure 6.5). Examples are sodium iodide (NaI) and bismuth germanate (BGO). These materials are used for radiation monitoring, in research, and in medical imaging equipment.

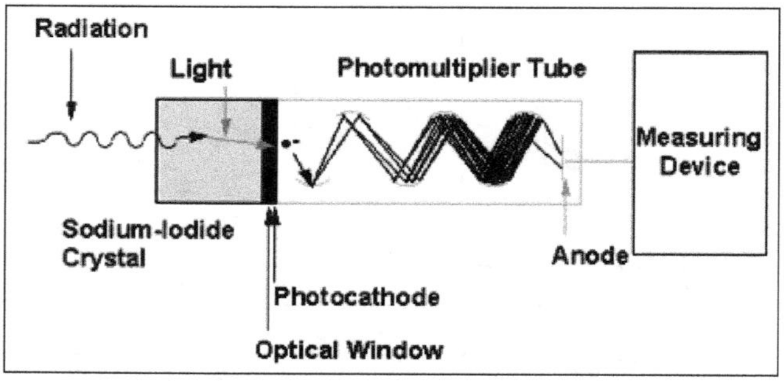

Figure 6.5: A Scintillation Detector.

Scintillation counters are made from materials that emit light when charged particles move through them. To detect these events and to gain information about the radiation, some means of detecting the light must be used. One of the first scintillation detectors was a glass screen coated with zinc sulfide. This sort of detector was used by New Zealand-British physicist Ernest Rutherford in the early versions of his classic experiment in which he discovered the nucleus of the atom by scattering alpha particles from heavy atoms such as gold. The scattered alpha particles hit the scintillating screen. Experimenters in a darkened room using only the human eye observed the small flashes that were produced.

The modern scintillation counter usually uses what is called a photo multiplier tube to detect the light. Light incident on the photo-cathode of such a tube is converted into an electrical signal and amplified millions of times after which it can be sent to appropriate counters. Physicists working at particle accelerators often use transparent plastic materials like Lucite or plexiglass to which materials are added to make them scintillate. These plastic scintillators can be cut to convenient shapes, mounted on a photomultiplier tube and placed in particle beams to provide a very fast signal when charged particles pass through them.

Since, the intensity of the light pulse emitted by a scintillator is proportional to the energy of the absorbed radiation, the latter can be determined by measuring the pulse height spectrum. This is called spectroscopy. To detect nuclear radiation with a certain efficiency, the dimension of the scintillator should be chosen such that the desired fraction of the radiation is absorbed. For penetrating radiation, such as g-rays, a material with a high density is required. Furthermore, the light pulses produced somewhere in the scintillator must pass the material to reach the light detector. This imposes constraints on the optical transparency of the scintillation material.

When increasing the diameter of the scintillator, the solid angle under which the detector "sees", the source increases. This increases detection efficiency. Ultimate detection efficiency is obtained with so-called "well counters" where the sample is placed inside a well in the actual scintillation crystal.

The thickness of the scintillator is the other important factor that determines detection efficiency. For electromagnetic radiation, the required thickness to stop, say, 90 % of the incoming radiation depends on the X-ray or g-ray energy. For electrons (*e.g.* b-particles) the same is true but different dependencies apply. For larger particles (*e.g.* a-particles or heavy ions) a very thin layer of material already stops 100 % of the radiation.

The thickness of a scintillator can be used to create a selected sensitivity of the detector for a distinct type or energy of radiation. Thin (*e.g.* 1 mm thick) scintillation crystals have a good sensitivity for low energy X-rays but are almost insensitive to higher energy background radiation. Large volume scintillation crystals with relatively thick entrance windows do not detect low energy X-rays but measure high energy gamma-rays efficiently.

A very useful scintillation detector, particularly for the measurement of gamma-rays, utilizes a transparent crystal of NaI (sodium iodide) mounted on a photomultiplier tube. These crystals are particularly useful because charged particles produce in them an amount of light directly proportional to their energy over a wide range.

6.3.2 Neutron Detectors

Since neutrons are uncharged, their detection must depend on an interaction with matter that produces energetic charged particles. There are several nuclear reactions initiated by neutrons that result in charged particles. One of the most useful for slow neutrons is the reaction in which a neutron is incident on a boron nucleus. This reaction produces a lithium nucleus and an alpha particle, both of which are rapidly moving. Note that it is the boron isotope of mass 10, with a natural abundance of about 20%, that is required for this reaction and that the alpha particle is simply the nucleus of the helium atom. The boron is usually incorporated in the gas molecule BF3(boron trifluoride) that can be used as the gas in a proportional counter, which is much like a Geiger counter. The difference is simply that the voltages used are lower so that the discharge does not spread disruptively along the entire central electrode with the result that the electrical signal coming from the tube is proportional to the number of ions produced.

The signals are much smaller than a Geiger tube and require more amplification but the signal produced by the lithium nucleus and alpha particle, both of which are heavily ionizing, is relatively large and easily distinguishable. For fast neutrons, the probability of this boron reaction becomes very low such that other methods are required. A useful technique is to use a proportional counter filled with hydrogen. Fast neutrons colliding with the protons in hydrogen produce energetic protons that produce a signal from the counter.

Neutrons are much harder to detect because they are not charged. They are detected by nuclear interactions that produce secondary charged particles. For example, boron trifluoride (BF3) counters make use of the 10B(n,a) 7Li reaction to detect neutrons. Often one uses a moderator, such as paraffin, to slow the neutrons and thus increase the detection efficiency. These detectors are used to monitor the neutron fluxes in the vicinity of a reactor or accelerator. Liquid scintillators can measure both neutrons and gamma-rays. By carefully measuring the shape of the electronic signal, scientists can distinguish between these two types of particles.

The most common neutron detectors are of the proportional gas type. Since neutrons themselves have no charge and are non-ionizing, they are not so easy to detect as X-rays. The method of detection relies on the absorption of the neutron by an atom with the simultaneous emission of a γ-ray photon, often referred to as an (n,γ) reaction. Since the absorbing material must absorb neutrons and be capable of existing in gaseous form, then the choice of materials is very limited. The most common is ^{3}He gas, which relies on the reaction:

$$^3\text{He} + {}^1\text{n} \rightarrow {}^4\text{He} + \gamma$$

This produces the stable isotope of helium together with a γ-ray of a specific energy. Another suitable gas is BF_3, which uses the absorption properties of the isotope^{10}B. For thermal wavelength neutrons the ^{3}He gas detector is considered best, but for long wavelength neutrons BF_3 is actually better, but is rarely used now due to the technical difficulties of handling a corrosive and toxic gas. Because the capture cross-section of gases is very small, ^{3}He detectors are usually filled above atmospheric pressure, *e.g.* 5 to 10 bar. In size, they are typically 10 to 15 cm long and 2 to 5 cm in diameter.

Recent developments by the detector group at the ILL, Grenoble, have led to the production of single-tube He-gas detectors with positional sensitivity. This is achieved by measuring the charge developed at both ends of a resistive anode wire, a larger magnitude of charge detected being related to a shorter distance travelled by the pulse along the wire. These will have huge potential for high-resolution angle-dispersive powder diffractometers since they will allow a greater solid angle to be measured without comprising instrumental resolution, the asymmetry due to the curved Debye-Scherrer diffractions being removed via software binning of the 2-dimensional data to the conventional 1-dimensional powder pattern.

Neutron multi-detectors are also ^{3}He gas based, though in addition they are also filled with a quenching gas such as a mixture of xenon (Xe) and methane (CH_4). The original multi-detectors consisted of an array of vertical wires at fixed intervals apart, usually chosen so that the interval corresponded to a nice angle in degrees, *e.g.* 0.1°. They are very large, can be both one- and two-dimensional, and can cover a huge area of solid angle. A recent development to improve the spacing interval has been the development of micro-strips which replace the conventional anode wires in these detectors.

Scintillation detectors have been developed using, for example, ^{6}Li or Gd as the absorbing atoms. The latter may be used in the mixed metal oxide-sulphide $Gd_2O_3.Gd_2S_3$, which is able to convert the γ-ray photon directly into a UV/visible photon. The latter is counted by using a conventional photomultiplier tube. With scintillation detectors, the neutron is detected in a smaller region of physical space compared to a gas detector due to the higher density of absorbing material. This has advantages for some applications. A potential disadvantage is that they may detect sample fluorescence due to (n,γ) reactions within the sample itself.

Thus one of the problems in detecting neutrons is the potential background from rogue γ-rays produced by neutrons from either the reactor or the pulsed source, plus those produced by reaction with a monochromator or even the sample. Therefore, the electronics controlling the detector must be capable of discriminating the energy (or emission time at pulsed sources) of the γ-ray so that only those produced in the detector are counted. In addition, the detector must be highly shielded from the general "cloud" of neutrons that exist around the diffractometer. The shielding is very thick: many centimetres of polythene are used to slow down any fast neutrons; the background neutrons then being absorbed by B_4C-doped plastic or rubber.

In a laboratory, the source is assumed to produce a constant flux of X-ray photons that does not vary with time. As discussed earlier, this is certainly not the case for synchrotron X-rays since the beam decays with time. In principle, the flux of neutrons from a reactor is constant with time, though in some cases it may not be, *e.g.* if the reactor has to run for a period at reduced power due to pump failure. Also at a pulsed source the intensity of the neutron beam is not constant in a short time interval, although averaged over a long time period, it is effectively constant.

Powder neutron diffractometers need to monitor the incident intensity of the radiation illuminating the sample. This is done using a monitor, which is basically a detector that works at very low efficiency. Monitors transmit most of the neutrons passing through them while absorbing a very small percentage, *e.g.* 0.1%. Neutron monitors are often fabricated like ^{3}He detectors, but with very thin aluminium windows, since the internal gas is at either at low pressure or is diluted to atmospheric with ^{4}He.

6.3.3 Semiconductor Detectors

Devices realized adopting semiconductor materials exhibit crystal flaws which can be caused by dangling bonds at interfaces or by the presence of impurities in the substrate. The presence of these defect centres, or traps, in semiconductor substrates may significantly influence the device electrical characteristics. Trap centres, whose associated energy lies in the forbidden gap, exchange charges with semiconductor conduction and valence bands through the emission and recombination of electrons. Therefore, trap centres change the space charge density in the semiconductor bulk and consequently, they both influence the recombination statistics and induce distortions in the electric field distribution. Therefore, in designing radiation detectors, close attention should be paid to semiconductor growth, device structure and technology processes.

As, both detectors and detection methods are currently field of developments and investigations, this review study considers, also, the physics principles of radiation interaction with matter as a general knowledge background needed to understand how radiation can be detected. Therefore, the second section is dealing with radiation matter interaction taking into account both directly and indirectly ionizing beams. For the detector performance evaluation, some characteristic parameters are introduced. In fact, energy resolution, detection efficiency and dead time are some of the detector properties. A particular attention has given to the discussion of semiconductor detectors.

Semiconductor detectors can be considered analogous to ionization chambers with the gas being replaced by the semiconductor material. The group 4 elements silicon and germanium are most widely used as semiconductor detectors when excellent energy resolution is required. Semiconductors have a small energy gap between their valence and conducting bands. At energies above absolute zero, thermal excitation will move some electrons from the valence band to the conductor band and holes or positive charges are left in the valence band. The quasi-free electrons in the conductor band move towards the anode if an electric field is applied. If ionizing radiation interacts with a semiconductor, electrons are transferred from the valence to the conducting band. The energy necessary to produce an ion pair is low (3–5 eV). Thus a large number of electrons, proportional to the absorbed radiation energy, can be collected at the anode by applying a high voltage of 1–2 kV to the detector, and statistical scatter is low. The current of the radiation-induced free electrons is superimposed by the temperature-dependent free charge density (leakage current) and electron recombination by acceptor-type impurities. Silicon cannot be produced at purity levels that allows electron recombination by acceptor-type impurities to be neglected. In Si(Li) detectors acceptor-type impurities are compensated for by drifting the silicon with lithium as donator.

In order to reduce the thermal charge carrier generation (noise) to acceptable levels Si(Li) detectors and Ge detectors are operated at liquid nitrogen (LN2) temperatures. LN-cooled detectors are mounted in a vacuum chamber, which is inserted in or attached to an LN2 Dewar flask. The detector is in thermal contact with the LN2 and the radiation enters the detector housing via a thin Be window. In mobile systems, and for applications requiring long-term unattended operation, detectors cooled by Joule–Thomson coolers, He-cycle refrigerators or Peltier elements are used, but all of them can show degraded energy resolution, with the Peltier exhibiting the most (~20%). Extremely compact room temperature detectors using semiconductors with higher energy gaps (CdZnTe, HgI_2, GaAs) have been developed in the 1990s. Energy resolution of these detectors is poorer than for Si(Li) detectors but better than with proportional or scintillation detectors.

A semiconductor detector is a device that uses a semiconductor (usually silicon or germanium) to detect traversing charged particles or the absorption of photons (Figure 6.6). In the field of particle physics, these detectors are usually known as silicon detectors. When their sensitive structures are based on a single diode, they are called semiconductor diode detectors. When they contain many diodes with different functions, the more general term semiconductor detector is used. Semiconductor detectors have found broad application during recent decades, in particular for gamma and X-ray spectrometry and as particle detectors.

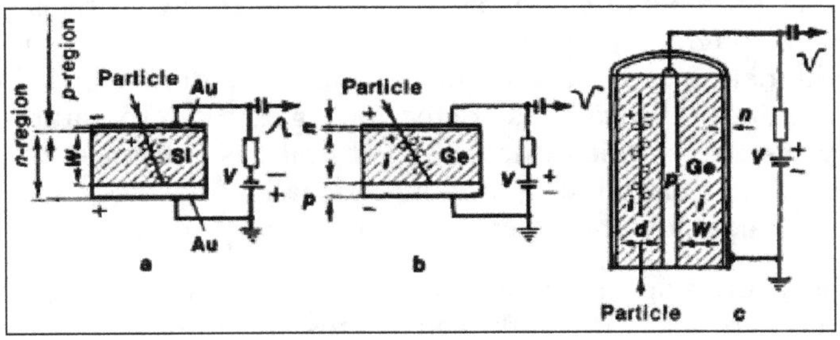

Figure 6.6: Semiconductor Detector.

In these detectors, radiation is measured by means of the number of charge carriers set free in the detector, which is arranged between two electrodes. Ionizing radiation produces free electrons and holes. The number of electron-hole pairs depends on the energy transmitted by the radiation to the semiconductor. As a result, a certain number of electrons are transferred from the valence band to the conduction band, and an equivalent number of holes are created in the valence band. Under the influence of an electric field, electrons as well as holes travel to the electrodes, where they give rise to a pulse that can be measured in an outer circuit. The holes travel into the opposite direction and can also be measured.

The energy required for production of electron-hole-pairs is very low compared to the energy required for production of paired ions in a gas detector. Consequently, in semiconductor detectors the statistical variation of the pulse height is smaller and the energy resolution is higher. As the electrons travel fast, the time resolution is also very good. Compared with gaseous ionization detectors, the density of a semiconductor detector is very high, and charged particles of high energy can give off their energy in a semiconductor of relatively small dimensions.

The disadvantages of semiconductor detectors are a low detection efficiency for high-energy γ-quanta, a deterioration of resolution for loads of more than 10^4 particles per second, and a limited service life of the detector at high doses of radiation because of the accumulation of radiation defects. The small dimensions of available single crystals (diameter ~ 3 cm, volume ~ 100 cm^3) limit the use of semiconductor detectors in a number of areas.

The next stage in the development of semiconductor detectors is connected with the production of super pure large-sized semiconductor single crystals and the possibility of using GaAs, SiC, and CdTe. Extensive use is made of semiconductor detectors in nuclear physics, the physics of elementary particles, chemistry, geology, medicine, and industry.

6.3.4 Gas Detectors

Gas detectors measure and indicate the concentration of certain gases in an air via different technologies. Typically employed to prevent toxic exposure and fire, gas detectors are often battery operated devices used for safety purposes. They are manufactured as portable or stationary (fixed) units and work by signifying high levels of gases through a series of audible or visible indicators, such as alarms, lights or a combination of signals. While many of the older, standard gas detector units were originally fabricated to detect one gas, modern multifunctional or multi-gas devices are capable of detecting several gases at once. Some detectors may be utilized as individual units to monitor small workspace areas, or units can be combined or linked together to create a protection system.

As detectors measure a specified gas concentration, the sensor response serves as the reference point or scale. When the sensors response surpasses a certain pre-set level, an alarm will activate to warn the user. There are various types of detectors available and the majority serves the same function: to monitor and warn of a dangerous gas level. However, when considering what type of detector to install, it is helpful to consider the different sensor technologies.

Technologies

Gas detectors are categorized by the type of gas they detect: combustible or toxic. Within this broad categorization, they are further defined by the technology they use: catalytic and infrared sensors detect combustible gases and electrochemical and metal oxide semiconductor technologies generally detect toxic gases.

Measurement of Toxic Gases

Electrochemical sensors or cells are most commonly used in the detection of toxic gases like carbon monoxide, chlorine and nitrogen oxides. They function via electrodes signals when a gas is detected. Generally, these types of detectors are highly sensitive and give off warning signals via electrical currents. Various manufacturers produce these detectors with a digital display.

Metal Oxide Semiconductors, or MOS, are also used for detecting toxic gases (commonly carbon monoxide) and work via a gas sensitive film that is composed of tin or tungsten oxides. The sensitive film reacts with gases, triggering the device when toxic levels are present. Generally, metal oxide sensors are considered efficient due their ability to operate in low-humidity ranges. In addition, they are able to detect a range of gases, including combustibles.

Measurement of Combustible Gases

Catalytic sensors represent a large number of gas detector devices that are manufactured today. This technology is used to detect combustible gases such as hydrocarbon, and works via catalytic oxidation. The sensors of this type of detector are typically constructed from a platinum treated wire coil. As a combustible gas comes into contact with the catalytic surface, it is oxidized and the wiring resistance is changed by heat that is released. A bridge circuit is typically used to indicate the resistance change.

Infrared sensors or IR detectors work via a system of transmitters and receivers to detect combustible gases, specifically hydrocarbon vapors. Typically, the transmitters are light sources and receivers are light detectors. If a gas is present in the optical path, it will interfere with the power of the light transmission between the transmitter and receiver. The altered state of light determines if and what type of gas is present.

Applications

Although detectors are an essential application for home and commercial safety, they are also employed in numerous industrial industries. Gas detectors are used in welding shops to detect combustibles and toxics and in nuclear plants, to detect combustibles. They are also commonly used to detect hazardous vapors in wastewater treatment plants.

Gas detectors are very efficient in confined spaces where there is no continuous employee occupancy. Such spaces include tanks, pits, vessels and storage bins. Detectors may also be placed at a site to detect toxins prior to occupant entry.

6.4 Detectors Used in Radiation Protection

6.4.1 Ionization Chamber

Ionization chamber is a device used for detection and measurement of ionizing radiation. It consists basically of a sealed chamber containing a gas and two electrodes between which a voltage is maintained by an external circuit. When ionizing radiation, *e.g.*, a photon, enters the chamber (through a foil-covered window), it ionizes one or more gas molecules. The ions are attracted to the oppositely charged electrodes; their presence causes a momentary drop in the voltage, which is recorded by the external circuit. The observed voltage drop helps identify the radiation because it depends on the degree of ionization, which in turn depends on the charge, mass, and speed of the photon.

A simple type of pulse ionization chamber has two conducting electrodes in a container filled with gas as shown. A battery, or other power supply, maintains an electric field between the positive anode and the negative cathode. When ionizing radiation penetrates the gas in the chamber—entering, for example, through a thin gas-tight window—this radiation liberates electrons from the gas atoms leaving positively charged ions. The electric field present in the gas sweeps these electrons and ions out of the gas, the electrons going to the anode and the positive ions to the cathode (Figure 6.7).

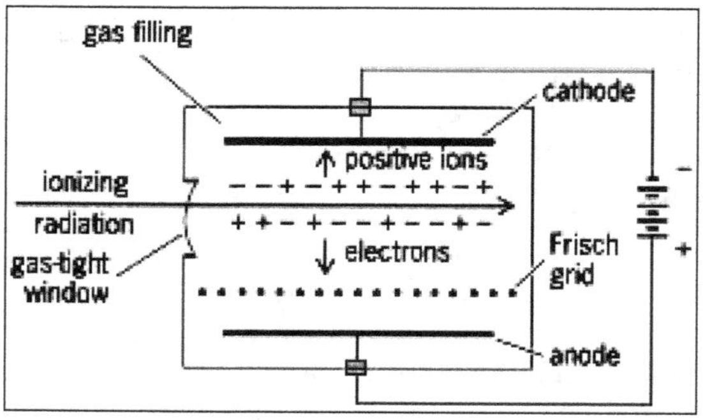

Figure 6.7: Pulse Ionization Chamber

In a chamber, such as that represented in the illustration, the current begins to flow as soon as the electrons and ions begin to separate under the influence of the applied electric field. The time it takes for the full current pulse to be observed depends on the drift velocity of the electrons and ions in the gas. Because the ions are thousands of times more massive than the electrons, the electrons always travel several orders of magnitude faster than the ions. As a result, virtually all pulse ionization chambers make use of only the relatively fast electron signal.

One of the most important uses of an ionization chamber is to measure the total energy of a particle or, if the particle does not stop in the ionization chamber, the energy lost by the particle in the chamber. In addition to energy information, ionization chambers are now routinely built to give information about the position within the gas volume where the initial ionization event occurred. This information can be important not only in experiments in nuclear and high-energy physics where these position-sensitive detectors were first developed, but also in medical and industrial applications.

Foremost among the other applications is the use of gas ionization chambers for radiation monitoring. Portable instruments of this type usually employ a detector containing approximately 60 in.3 (1 liter) of gas, and operate by integrating the current produced by the ambient radiation. Another application of ionization chambers is the use of air-filled chambers as domestic fire alarms. Yet another development in ion chamber usage is that of two-dimensional imaging in X-ray medical applications to replace the use of photographic plates.

Applications

Nuclear Industry

Ionization chambers are widely used in the nuclear industry as they provide an output that is proportional to radiation dose They find wide use in situations where a constant high dose rate is being measured as they have a greater operating lifetime than standard Geiger-Müller tubes, which suffer from gas break down and are generally limited to a life of about 1011 count events. Additionally, the Geiger-Müller tube cannot operate above about 104 counts per second, due to dead time effects, whereas there is no similar limitation on the ion chamber.

6.4.2 Smoke Detectors

The ionization chamber has found wide and beneficial use in smoke detectors. In a smoke detector, ambient air is allowed to freely enter the ionization chamber. The chamber contains a small amount of americium-241, which is an emitter of alpha particles which produce a constant ion current. If smoke enters the detector, it disrupts this current because ions strike smoke particles and are neutralized. This drop in current triggers the alarm. The detector also has a reference chamber which is sealed but is ionized in the same way. Comparison of the ion currents in the two chambers allows compensation for changes due to air pressure, temperature, or the ageing of the source.

Medical Radiation Measurement

In medical physics and radiotherapy, ionization chambers are used to ensure that the dose delivered from a therapy unit or radio-pharmaceutical is what is intended. The devices used for radiotherapy are called "reference dosimeters", while those used for radio-pharmaceuticals are called radioisotope dose calibrators - an inexact name for radionuclide radioactivity calibrators, which are used for measurement of radioactivity but not absorbed dose.

6.4.3 Proportional Counters and Geiger Muller Counters

Proportional Counters

The proportional counter is a type of gaseous ionization detector device used to measure particles of ionizing radiation. The key feature is its ability to measure the energy of incident radiation, by producing a detector output that is proportional to the radiation energy; hence the detector's name. It is widely used where energy levels of incident radiation must be known, such as in the discrimination between alpha and beta particles, or accurate measurement of X-ray radiation dose.

A proportional counter uses a combination of the mechanisms of a Geiger–Müller tube and an ionization chamber, and operates in an intermediate voltage region between these. The accompanying plot shows the proportional counter operating voltage region for a co-axial cylinder arrangement.

In a proportional counter the fill gas of the chamber is an inert gas which is ionised by incident radiation, and a quench gas to ensure each pulse discharge terminates; a common mixture is 90% argon, 10% methane, known as P-10. An ionising particle entering the gas collides with an atom of the inert gas and ionises it to produce an electron and a positively charged ion, commonly known as an "ion pair". As the ionising particle travels through the chamber it leaves a trail of ion pairs along its trajectory, the number of which is proportional to the energy of the particle if it is fully stopped within the gas. Typically a 1 MeV stopped particle will create about 30,000 ion pairs.

The chamber geometry and the applied voltage is such that in most of the chamber the electric field strength is low and the chamber acts as an ion chamber. However, the field is strong enough to prevent re-combination of the ion pairs and causes positive ions to drift towards the cathode and electrons towards the anode. This is the "ion drift" region. In the immediate vicinity of the anode wire, the field strength becomes large enough to produce Townsend avalanches. This avalanche region occurs only fractions of a millimeter from the anode wire, which itself is of a very small diameter. The purpose of this is to use the multiplication effect of the avalanche produced by each ion pair. This is the "avalanche" region.

A key design goal is that each original ionising event due to incident radiation produces only one avalanche. This is to ensure proportionality between the number of original events and the total ion current. For this reason the applied voltage, the geometry of the chamber and the diameter of the anode wire are critical to ensure proportional operation. If avalanches start to self-multiply due to UV photons as they do in a Geiger–Muller tube, then the counter enters a region of "limited proportionality" until at a higher applied voltage the Geiger discharge mechanism occurs with complete ionisation of the gas enveloping the anode wire and consequent loss of particle energy information.

6.4.4 Geiger Muller Counters

The Geiger-Müller counter operates in the Geiger-Müller region, in which collected charge is independent of initiating ionization and the rate of particle collection reaches a plateau. This type of detector is effective for counting lightly ionizing radiation, such as beta particles and gamma-rays. A Geiger-Müller tube usually incorporates a fine wire electrode and operates at voltages significantly higher than those of ionization chambers or proportional counters. Geiger-Müller counters normally cannot identify the specific type of radiation emitted or its energy level.

A Geiger counter is a metal cylinder filled with low-pressure gas sealed in by a plastic or ceramic window at one end. Running down the center of the tube there's a thin metal wire made of tungsten. The wire is connected to a high, positive voltage so there's a strong electric field between it and the outside tube.

Geiger-Muller (G-M) Counters produce a large voltage pulse because the numbers of ion pairs produced are greatly amplified, by the phenomenon known as gas multiplication. Gas multiplication is induced by a strong applied electric field.

After creation by radiation interaction, both electrons and ions make many collisions with neutral gas molecules until collected. When the electric field is strong enough, electrons get high kinetic energies between collisions and eventually can ionize neutral molecules as illustrated in Figures 6.8 while ions have low mobility and they attain very little energy between collisions.

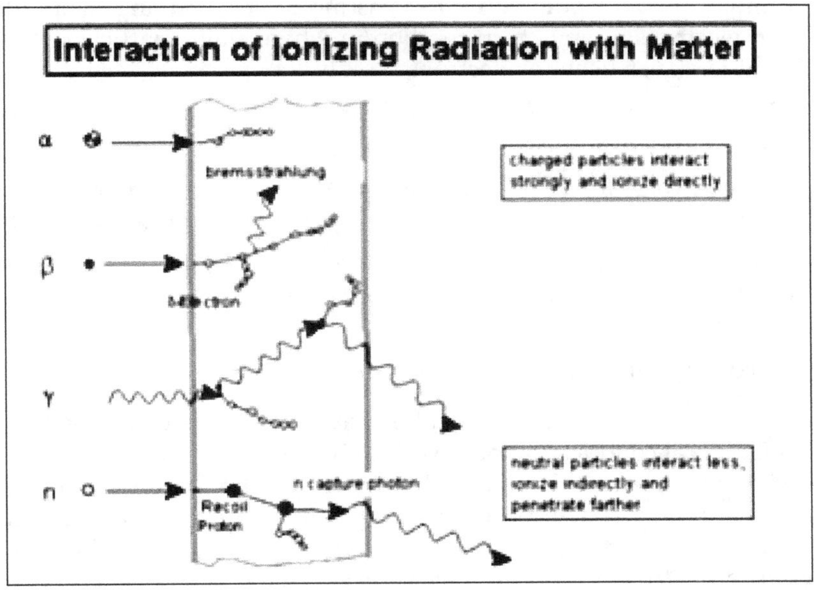

Figure 6.8: Interaction of Ionization Radiation with Matter.

The two electrons can then be accelerated and further ionization's can be caused. The process takes the form of a cascade. The electric field must be stronger than the threshold. The fractional increase in the number of electrons per unit path length is governed by the Townsend Equation:

$$dn/n = \alpha dx$$

Observation

When Geiger counter detects some radiation:

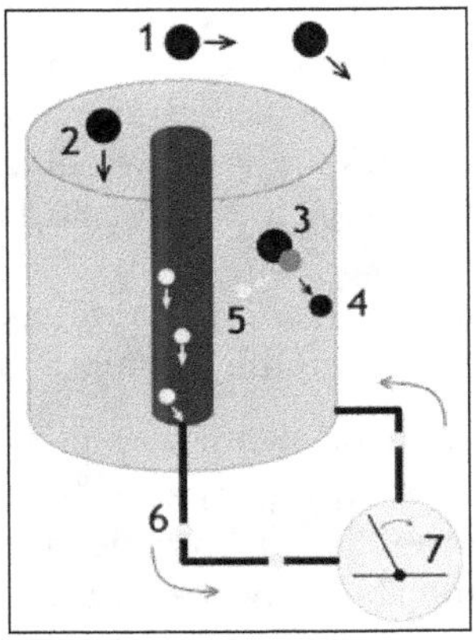

- Dark Blue Radiation will move about randomly outside the detector tube.
- Some of radiation will enter the window at the end of the tube.
- When dark blue radiation collides with gas molecules in the tube, it causes ionization: some of the gas molecules are turned into positive ions and electrons.
- The positive ions are attracted to the outside of the tube.

- The electrons are attracted to a metal wire running down the inside of tube maintained at a high positive voltage. As electrons head for wire, some of them collide with other gas molecules, splitting them into ions and electrons giving chain reaction in where single particle of radiation produces avalanches of electrons rapidly.
- Many electrons travel down the wire making a burst of current in a circuit connected to it.
- The electrons make a meter needle deflect and, if a loudspeaker is connected, you can hear a loud click every time particles are detected.
- Before the counter can detect any more radiation, it needs to be restored to its original state through a process called quenching, which cancels out the effects of the Geiger discharge.

6.4.5 Semiconductor Detectors/Counters

Semiconductors are materials that neither have enough free charge carriers to behave as electrical conductors nor have a high resistivity as that of the electrical insulators. As mentioned before, the solid crystal has three energy bands that are valence band, conduction band and forbidden band. For the electrical conductors, the width of the forbidden energy band is very small that allows the movement of the valence electrons to the conduction band, under the effect of any electric field strength higher than zero, where the electrons can move freely in the crystal lattice and carry electric current. For the electrical insulators, the width of the forbidden energy band is big (about 10 eV) enough to prevent the movement of the valence electrons to the conduction band, which is completely empty. For the semiconductor materials, the forbidden energy band is relatively narrow to prevent the movement of the electron to conduction band at low temperature *i.e.* the conductivity of the semiconductors is zero. As temperature increases, some electrons gain enough energy to cross the forbidden band to the conduction band where electrons can carry the electric current under the influence of an electric field in the same way as the conductors.

Following the discovery of semiconductor devices that could perform the circuit functions of thermionic valves and photoelectric cells - the transistors and photo resistors now so widely used, as there has been considerable success in the last few years in the development of semiconductor devices which can be used for the detection of ionizing radiation. These detectors have some exceptionally valuable features, and it can already be said that their introduction has very considerably advanced the technique of radiation measurement.

Semiconductor crystal as a detector material should have the capability of supporting large electric field gradients, high resistivity, and exhibits long life and mobility for both electrons and holes. If the mobility is too small and life time is too short, most electrons and holes will be trapped at crystal lattice imperfections or recombine before they can be collected. The group IV elements silicon and germanium are the most widely used semiconductor crystals as radiation detectors. Some of the key characteristics of various semiconductors as radiation detector materials. The conductivity of semiconductors increases with increasing the concentration of impurities, which create new energy levels that facilitate the movement of the carrier within the crystal. The ideal semiconductor material is called "intrinsic" or "low effective impurity" material that is produced by a process, called "doping", involves the addition of an impurity to reduce the charge carrier concentration, *i.e.*, adding an electron accepting impurity to compensate for electron donor impurities. Although doping increases the resistivity of the material it also increases the probability of electron-hole trapping or recombination. Prior to the mid-1970's, the required purity level of Si and Ge could be achieved only by, lithium-ion drifting process, counter-doping P-type (electron acceptor) crystals with the N-type (electron donor) impurity to produce Ge(Li) and Si(Li) crystals. Since 1976, sufficient pure Ge, hyper pure, crystal have been available but doping process is still widely used in the production of Si(Li) X-ray detectors.

Semiconductors have four valence electrons in the upper energy level of the valence band, if they are doped with atoms, as crystal impurity, with three valence electrons such as gallium, positive holes will be existed in the crystal that is known as P-type crystal and the holes are the major current carrier. While, if they are doped with atoms with five valence electrons such as arsenic, an excess electrons will be existed in the crystal that is known as N-type crystal and the electrons are the major current carrier. Semiconductors have a P-N diode structure and radiation detection is based on the favorable properties of the intrinsic region, the region near the junction between N- and P-type semiconductor materials, which is created by depletion of charge carriers. The depletion region is the sensitive volume of semiconductor detectors where the ionizing radiation interacts and the dissipated energy will produce electron-hole pairs directly or indirectly in the same way like gas filled detectors. Electron-hole pairs are swept to the P and N regions. The produced charge is linearly correlated to the energy deposited in the detector. Semiconductors might be considered 'as solid state ionization chambers' with several advantages over gas devices. Unbiased P-N junction can act like a detector but only with a very poor performance because the depletion region thickness is quite small, the junction capacitance is high, and the spontaneously electric field strength across the junction is low and not enough to collect the induced charge carriers that could be lost due to trapping and recombination. The performance of the P-N junction as a practically used radiation detector is improved by applying an external voltage that leads the junction to be reversed biased. As the applied voltage increases, the width of the depletion region and the sensitive volume are increased and the performance of the detection is improved. The applied voltage should be kept below the breakdown voltage of the detector to avoid a catastrophic deterioration of detector properties.

Semiconductor detectors are special ionization chambers, where the ionization occurs in solid, semiconductor material. (Instead of gas the ionization works is solid material.). Because of the narrow energy band gap, 0.74 eV for germanium and 1.12 eV for silicon, semiconductor detectors are thermally sensitive. Practically both Ge and Si photon detector are cooled down with liquid nitrogen during operation to reduce the thermal charge carrier generation (noise) to acceptable limit where the reverse leakage currents are in the range of 10-9 -10-12 amperes at liquid nitrogen temperature (77^0K). The narrow energy band gap of the semiconductor materials lets the energy required to produce an electron-hole pair is a tenth that required to produce an electron-hole pair in a gas. This gives them the advantage of better energy resolution over the gas filled and scintillation detectors where the increase in the number of charge carrier in semiconductor detector leads to improved statistics and better energy resolution. The excellent energy resolution of semiconductor detectors is due to the much larger number of charge carriers per pulse, *i.e.* they produce much larger number of charge carrier for a given incident radiation than is possible with any other detector type.

Parameters

- From the beginning of 1960′s
- Small size
- Good energy distribution capability
- It is nicely used for radiating surfaces with small radiation
- Silicon, Germanium are mostly well known materials

In principle, a semiconductor detector can be regarded as an ionization chamber in which the sensitive volume is a solid instead of a gas: it consists basically of a semiconductor wafer with an electrode on each side. An incident particle or quantum produces a certain ionization charge in the semiconductor, just as it would in a gas, and this gives a pulse of current in an external circuit. This in turn causes a voltage pulse to appear across a resistance included in the circuit.

The great attraction of using a solid as the detection medium is its higher absorption, which enables even fast particles to be stopped in a relatively thin layer, making it possible to measure their energy with an instrument of small dimensions. An incidental but extremely important advantage is that a semiconductor counter can be very much faster than a gas filled ionization chamber.

Features

- Solid state is denser material than gas whereas solid state needs weaker radiation to start the ionization
- Taking same irradiation, in solids more charged particles can be generated, than in gas ionization chamber.
- In case of solid semiconductor the measurement is more stable, errors are lower (lower error bars during the measurements)
- In case of semiconductors the energy resolution is better than in gases
- Specific conductivity is better for semiconductors

Application

- Measurement on heavy charged particles
- Detecting electrons
- Detecting gamma-radiations
- Observing neutrons
- Germanium detector

Germanium detector is made of hyper pure germanium (HPGe) crystal that is mounted in a vacuum chamber to be protected from moisture and considerable contaminant, and is cooled by liquid nitrogen cryostat inserted in a liquid nitrogen Dewar to reduce the leakage current to acceptable level. The preamplifier is located near the detector as part of the cryostat package to reduce electronic noise. A cross section view of a typical HPGe detector with the liquid nitrogen cryostat is shown in Figure 7. There are different types of germanium detector such as coaxial, planar and well type. Their geometry and construction features are shown in Figure 8. The geometry and construction features of the detector affect their detection features such as detection efficiency for gamma and X-rays, energy ranges and energy resolution. The variation of detector efficiency and energy resolution as a function of incident radiation energy for the different detector types are shown in Figures 9 and 10 respectively. Coaxial P-type germanium detectors are used for gamma-ray with energy range from 100 KeV to about 10 MeV and cannot be considered for low energy gamma and X-rays because they cannot penetrate the detector aluminum window and high energy gamma-rays that might pass through the sensitive volume without interaction. For X-ray spectroscopy, N-type and planar germanium detectors can be used because of the thin beryllium entrance windows. At low energies, detector efficiency is a function of the cross sectional area and windows' thickness while at higher energies, total active detector volume more or less determines counting efficiency. Coaxial germanium detectors are specified in term of their relative full-energy peak efficiency compared to that of he 3 in. x 3 in. NaI(Tl) scintillation detector at a detector to source distance of 25 cm. Germanium detector of greater than 100% relative efficiency have been fabricated.

Silicon Detectors

(a) Si(Li) X-ray detectors; P-type silicon crystal is doped with lithium atoms, lithium drifting process, to produce a very stable Si(Li) crystal that can be stored in room temperature and cold only during operation as X-rays detector. It is only useful for X-rays detection in the energy range between 4 and 50 keV. Signal to noise ratio is very significant parameter that affects the resolution and the performance of any radiation detector especially in low level radioactivity measurement. It is particularly critical problem with silicon detector because the energy deposited in silicon by X-rays and the produced electric signals are small. That is why the proper electronics should be used to reduce the noise and to amplify the signal. Both Si(Li) and planar germanium detector can be used for X-ray spectroscopy but the Si(Li) detector is better because its escape peaks are very low. The planar germanium detector can be very useful in high X-ray and very low energy gamma-rays measurements. X-rays interact with the silicon mainly by photoelectric effect and the spectrum looks simpler than that produced by using germanium detectors.

Si(Li) detector window is usually made of low Z (atomic number) materials such as beryllium, which are then enough for X-ray enter and strong enough to hold vacuum. In addition to the detector window thickness, X-ray energy will affect the detection efficiency where the efficiency increases with the increasing X-rays energy to reach the maximum and then decreases with increasing energy because at the high energy the X-rays might pass the active volume without interaction.

(b) Silicon charged particle detectors; they are also known as silicon diode detectors or surface barrier (SSB) detectors and the modern version are known as passivated implanted planar silicon (PIPS) detectors. They have become the detectors of choice of the majority of heavy charged particles measurements including alpha particles and fission fragments spectroscopy. They have a P-N structure in which a depletion region is formed by applying reverse bias. In the SSB detector, a surface barrier junction is formed by oxidizing the surface of the N-type silicon. Electric contact is made to the P-type oxidized surface by a thin layer of gold and to the N-type surface by a layer of aluminum. The intrinsic, depletion, region is the sensitive volume of the detector and is formed due to the migration of electron toward the P-type region and the hole toward the N-type region. The width of the depletion region increases with increasing the bias voltage and can extended to the limit of the breakdown voltage. The resistivity of the silicon must be high enough to allow a large enough depletion region at a moderate bias voltage. PIPS detectors employ implanted rather surface barrier contacts and are therefore more rugged and reliable than the SSB detectors. Detectors are generally available with depletion layer depths of 100 -700 μ m and active areas of 25-5000 mm^2.

6.4.6 Neutron Dosimeters

In comparison to gamma dosimetry, neutron dosimetry is still far less established. For the moment no neutron personal dosemeter is available that measures the neutron personal dose equivalent within a 50% uncertainty level. This lack of precision is mainly caused by the fact that the neutron dose is strongly energy and angular dependent. Moreover neutron energies in typical workplace fields vary with at least 10 orders of magnitude, demanding challenging requirements on neutron dosemeters and detectors.

Neutron dosimetry is of practical relevance in a wide energy range of at least ten orders of magnitude. However, the dose quantities used in radiation protection are not directly measurable. This is mainly due to the variable quality of neutron radiation, expressed by radiation weighting factors or quality factors. In spite of considerable progress in neutron dosimetry there is no dosemeter which is capable of measuring neutron doses independently of the neutron spectrum with adequate accuracy.

Complete neutron field characterization in terms of energy and directional distribution is therefore indispensable for a reliable estimation of the neutron dose equivalent.

Activities

Several measurement campaigns for neutron field characterization are being organized in several locations where the neutron dose can make a significant contribution to the occupationally received dose. Such locations can be found inside the containment structure of nuclear reactors and in the near environment of particle accelerators. Commercial devices are used for determination of the energy distribution and the total neutron fluency.

In case of a criticality accident, occupationally exposed workers can be the victim of severe neutron exposure. In such cases, specialized neutron dosimetry techniques are necessary, which differ markedly from those used in routine radiological protection. Therefore SCK•CEN has developed a system for criticality dosimetry based on activation detectors.

Applications

One of the most important applications of the research performed in neutron dosimetry is the establishment of local correction factors for neutron personal dosemeters. When neutron field characterization is performed in terms of energy and directional distribution, the neutron personal dose equivalent Hp(10) can be calculated. This allows us to compare the calculated value with the measured values by personal dosemeters and to propose site specific correction coefficients improving routine personal dosimetry.

Preparedness in the case of the event of a criticality accident. A big advantage of the current system based on activation detectors is that the neutron dosemeters do not require any maintenance or periodical read-out. Because criticality accidents occur, luckily, very rarely, therefore, a yearly test is performed in a well quantified neutron field in order to keep the knowledge and procedures up to date.

Personal Dosimeters

Personal dosimetry is a well-established field, but still constantly new developments are introduced. SCK•CEN's Radiation Protection Dosimetry and Calibration expert group is testing and improving many new techniques, and introduces them in practical applications.

The research in personal dosimetry helps in constantly upgrading the performance of the service.

Personal dosimeters are assigned to individuals based upon regulatory requirements and their potential occupational exposure to penetrating radiation. Dosimeters are not effective for measurement of exposure from H-3, C-14, P-33, S-35, and Ni-63.

6.5 Radioactivity Analysis

Radioactivity analysis of food and environmental samples is a very significant task in any radio ecological study and the accuracy of the measured natural and artificial radio-nuclides concentration affects the study conclusions and the consequence measures for environmental and human protection.

Radio-ecology studies include the following main tasks:

- Sampling planning and samples collection,
- Samples preparation,
- Radioactivity analysis,
- Quality control measures,
- Data evaluation,
- Study conclusions and measures.

In principle, the sample preparation is a process in which samples are taken through several steps to reach the most appropriate form for radioactivity measurement. It serves to preserve the sample and keep it in more homogenize form, *i.e.* includes drying, crushing homogenization and sieving the soil and sediment samples. Sample preparation serves also to increase the counting efficiency and sometime it is essential for performing the radioactivity analysis itself such as for alpha emitters analysis (*e.g.* uranium isotopes, plutonium isotopes and ^{210}Po) where the source sample should be a thin layer form to reduce the self-absorption of alpha particle and the counting process is impossible without source sample chemical preparation.

Radioactivity analysis process includes source sample preparation, choosing the counting techniques that fulfill the analysis' objectives, counting system adjustments such as reducing the electronic noise and the background counts, energy and efficiency calibration, MDA calculation, radioactivity measurement, and calculation of radioactivity concentration and its associated error. Radioactivity measurements are classified into relative measurements where the standard sources are used for measurement system calibration and absolute measurements that do not need any standard sources for radioactivity analysis. Absolute measurement methods include the defined solid angle method, the 4 π counting method, the internal gas counting method and other methods based on the coincidence technique.

The mostly used radioactivity analysis methods for food and environmental samples include 2π alpha/beta gas flow counter, liquid scintillation spectrometer, gamma spectrometers, beta particle spectrometers and alpha particle spectrometers.

2π Alpha/Beta Counting with a Gas Flow Counters

Low background gas flow proportional counter is a very useful radioactivity counting system for environmental samples. Many laboratories are using total alpha/beta measurements as a rapid method to estimate the activities of alpha and beta radio-nuclides in water, air filters, soil, sludge, and waste water. The main purpose of gross alpha/beta measurement is to provide adequate information concerning the activity within samples, and, thus to determine if further detailed analysis are required. This counting technique is based on the capability of the proportional counters to differentiate between the ionization events due to alpha and beta particles that depend on the voltage bias applied on the counter. There are a variety of available counters that offer simultaneous separate alpha and beta measurement where for each sample there are two separate measuring channels for alpha and beta activities. The ultra-flat flow proportional counter tubes are located above the source samples planchets, as shown in Figure 16. To reduce the background, the measuring counters are covered at the top by a large-area flow counter tube acting as common guard counter, *i.e.* active shielding, in addition to a 10 cm thick lead, *i.e.* passive shielding, surrounds all counters.

Gross alpha/beta measurement techniques are applicable for alpha emitters having energies above 3 MeV and for beta emitters having energies above 0.1 MeV. The solid concentration in the source sample is a limiting factor in determining the sensitivity of the gas flow proportional counting because of self-absorption of alpha and/or beta particles by the sample itself. Because alpha and beta particles are attenuated or scattered by the sample solids, calibration curves for self-absorption and scattering must be established if there is mass in a sample. Sample density on the planchet area should not exceed 10 mg cm^{-2} for gross alpha, and 20 mg cm^{-2} for gross beta for a 2-inch diameter counting planchet. Moisture absorbed by the dried sample residue is an interferant because it enhances self-absorption.

The 2-in. diameter gas proportional detectors have a manufacturer measured alpha background of 0.05 counts min^{-1} and a beta background of 1.0 counts min^{-1}. An instrument with 1.0 counts min^{-1} beta background will have an LLD of 37 mBq for a 60-min count at the 95% confidence level, with 30% detection efficiency, whereas an instrument with an alpha background of 0.05 counts min^{-1} will have an LLD of 67 mBq for the same count time and detection efficiency.

When measuring alpha and beta particle activity using a gas-flow proportional system, counting at the alpha voltage plateau discriminates against beta particle activity, whereas counting at the beta voltage plateau is somewhat sensitive to the alpha particle activity present in the sample. This phenomenon is termed "cross talk" and is compensated for during instrument calibrations. The "cross talk" factors are determined using alpha ($^{238,239}Pu$ or ^{241}Am) and beta (^{90}Sr or ^{137}Cs) solutions of known activity concentrations counted over a range of absorber thicknesses (mg cm^{-2}) encountered during routine sampling and counting.

For gross alpha and gross beta measurements, detector must be calibrated to obtain the counting efficiency, *i.e.* the ratio of the counting rate to the disintegration rate. Alpha emitting radionuclide such as ^{241}Am, ^{230}Th or ^{238}Pu and beta emitting radionuclide such as $^{90}Sr/^{90}Y$ reference standard are prepared in the same samples geometries and weight ranges to be used for efficiency calibration where both self-absorption and backscattering of the alpha and beta particles are considered. Sample weight stability, which can be obtained by heating the sample to a red heat to convert the hygroscopic salts to oxides, is essential to ensure the accuracy of the self-absorption counting efficiency factor to be for the samples.[11] To overcome the possible uncertainty in the gross alpha/beta counting of soil samples due to the lack of well-characterized soil standards for calibrations and the unpredictable absorption effects that soil has on counting results. Burnett 1998 has proposed an alternative approach based on direct counting of pressed soil wafers to estimate total alpha/beta activities.

In addition to gross alpha/beta screening measurements, gas flow proportional counter is used also for analysis of some alpha emitting radio-nuclides such as ^{226}Ra and beta emitting radio-nuclides such as ^{210}Pb and $^{90}Sr/^{90}Y$ in different environmental samples after the proper radiochemical analysis to separate the radio-nuclides of interest in a pure and standard form source sample as an electroplating source or mostly as precipitation.

Liquid Scintillation Spectrometer

Since the early applications of internal sample liquid scintillation counting (LSC), the LSC systems have been designed mainly to detect low energy beta particles emitters (*e.g.*, ^{3}H, ^{14}C), their applications have been extend to detect high energy beta particles emitters (*e.g.*, ^{90}Y, ^{210}Bi), some of the new versions of LSCs are able to detect alpha particles emitters (*e.g.*, U isotopes, $^{239+240}Pu$) and known as LSC with alpha/beta discriminator. Samples containing mixtures of radio-nuclides that emit alpha and beta particles, conversion electrons or Auger electrons can be detected and quantified using variations of the LSC techniques. The liquid scintillations counter, liquid radiation detection media, play the role of an energy transducer, converting energy from nuclear decay into light. The light is sensed by photomultiplier tubes (PMT) and generates electric pulses, which are coincidence-gate discriminated, and analyzed by means of integrated electronic circuits and logic according to their timing and amplitude which is further recorded and displayed as counts, or count rate or energy spectrum. The liquid scintillation measurement method is quantitative technique to determine radioactivity, but not a qualitative one to find radiation energy or the sort of nuclide.

Liquid scintillation, known as LS cocktail, consists of solvent and organic scintillators that may be composed of one, two or three solutes. Usually the samples are incorporated in the LS cocktail to be assayed that results in either a homogeneous or heterogeneous sample depending on the solubility of the sample in the cocktail. Homogeneous sample counting is the preferred modality where the counting efficiency is 4π counting geometry and relatively high. Consequently, additives or solubilizers are often used to promote sample homogeneity. Since the sample comes into intimate contact with the scintillation components, it interferes with the energy transfer processes that known as quenching, and reduces the scintillation counting efficiency. Quenching processes are either chemical or color quenching, which reduce the maximum light output of the scintillation system. In heterogeneous sample counting, the sample to be assayed is either finely dispersed or suspended, or counted on a solid support by immersion in a liquid scintillation where the counting geometry may not be 4π geometry, *i.e.* relatively low counting efficiency and the quenching processes have only slightly effect or not at all the scintillation efficiency.

The three classes of liquid scintillations are;

Toluene- or Xylene-Based Scintillators

They are the most efficient light transducer, provided that their energy transfer process is not disturbed by the presence of impurities. Toluene-based scintillators are relatively stable and easy to prepare using scintillation grade chemicals.

P-Dioxane-Based Scintillators

They were introduced to meet the need for measuring aqueous samples. P-Dioxane is miscible with water but is an inefficient scintillator solvent. When supplemented with a high concentration (50-100 g/l) of naphthalene, a dioxane-based scintillator may attain efficiency about 70% of a toluene-based scintillator. They have obvious disadvantages such as its high cost and the solvent p-dioxane is readily oxidized during storage forming peroxides, which cause severe quenching and gives rise to intense chemiluminescence (chemical quenching) in the presence of alkaline tissue solubilizers.

Emulsion or Colloid Scintillators

They have a high capacity for holding aqueous samples, high counting efficiency and, the advantage of low cost and great flexibility. The emulsion scintillators are not stable counting systems, *i.e.* are not thermodynamically stable, and with slight changes in composition, the appearance and counting efficiency can undergo abrupt change.

Depending on the LS cocktail, the alpha detection efficiency is generally > 95%, whereas the beta detection efficiency is dependent on energy, spectral shape and cocktail. Typically, beta particles with maximum energies (E_{max}) > 0.250 MeV are detected with > 90% counting efficiency. Practically, the counting efficiency is reduced due to different interferences, *i.e.* quenching that result in photon quenching, chemical quenching and color quenching; these are processes in which the maximum photon yield is not achieved for a given radioactive source due to the sample inhomogeneity, adverse energy transfer, and non-transparency of the liquid scintillation solution to the photon emitted, respectively, therefore, quenching correction and efficiency determination are identical. Since samples vary in quench characteristics, their efficiency must be determined individually.

The most frequently used methods for quenching correction are:

Internal Standard Method

When properly carried out, it is the most accurate method for quenching correction in which a calibrated standard of the measured radionuclide is added to a sample that has been previously counted for efficiency determination.

$$Counting\ efficiency = \varepsilon = \frac{C_2 - C_1}{A}$$

Where:

C_1 = count rate of the sample in cps,

C_2 = count rate of the sample plus the added internal standard in cps,

A = calibrated activity of the added internal standard in Bq.

Quenched and Unquenched Standards Method

Series of liquid scintillation samples have the same calibrated standards activity of ^{3}H and ^{14}C and each sample has a successively increasing amounts of a quencher or non-radioactive equivalent of the material to be assayed. A relationship between the loss in count or counting efficiency and the quenching level can be computed and used for counting efficiency correction

Channels-Ratio Method

It is based on a downward shift of the pulse height spectrum of the radio-nuclides in the presence of a quencher. The extent of the shift is related to the counting efficiency of the liquid scintillation system.

External Standard Method

It is the use of Compton recoil electrons from high energy gamma-rays, photo-electrons from low energy gamma-rays (< 60 KeV), or X-rays, or a combination of these, for quenching correction. It is an integral part of all modern LS spectrometer. A correlation curve, calibration curve, between the count rate of the gamma-ray standard and the sample counting efficiency is obtained using a set of quenched standard.

The alpha emitting radio-nuclides can be measured by LS counter where the pulse-shape discrimination (PSD) is used for the characteristics of pulses from beta particles differ from those derived from alpha particles. Alpha emitters may be counted by direct incorporation of aqueous samples containing the nuclide in an emulsion-type scintillator or by extracting the nuclide of interest from the aqueous sample with an extractive scintillator. The LSC spectrum of electroplated $^{236+239+240}PU$, alpha emitting radio-nuclides, and ^{241}Pu, beta emitting radio-nuclides.

Gamma-rays Spectrometry

Gamma-rays measurement techniques are the backbone tool for radio-nuclides concentration estimation in environmental samples because most radio-nuclides emit gamma-ray, high penetration of gamma-ray permits comparatively simple source preparation, and gamma-ray spectrometry gives good selectivity in discriminating among radio-nuclides. The most important detection media for gamma-rays spectrometry are inorganic scintillator such as NaI(Tl), CsI(Na) and $CaI_2(Eu)$, semiconductors such as hyper pure germanium (HPGe). Practically the most often used gamma-ray detectors are NaI(Tl) and HPGe detectors.

The probability of the various types of interaction will depend upon the dimensions and the nature of the detecting medium and upon the energy of the incident photon. The total intrinsic detection efficiency is appreciably dependent upon the geometric arrangement. Photo-peak/total ratio or photo-fraction is relatively independent of geometry for a given detector. The selection of gamma-ray spectrometer is based on its photo-peak counting efficiency, energy resolution and costs. Gamma-ray spectrometry based on HpGe detector is preferred for the determination of radio-nuclides in food and environmental samples because of its higher resolving power, energy resolution = Full Width at Half Maximum (FWHM), of the HpGe detector than that of the NaI(Tl) detector; high energy resolution is essential for analysis of complex gamma spectra. NaI(Tl) scintillation is preferred when the high energy resolution is not essentially required, in case of measurement of few gamma emitters radio-nuclides that have no interference or overlapped photo-peak.

The energy resolution of NaI(Tl) crystal is about 6 % for the photo-peak of Cs-137 at 661.6 keV energy, *i.e.* it is about 40 keV and about 70 keV for the photo-peak of Co-60 at 1332 keV energy. While, the energy resolution of HpGe detector is about 1.9 keV for photo-peak of Co-60 at 1332 keV. The main advantages of NaI(Tl) are that they can be obtained in a larger sizes, they have a high efficiency and they cost less than semiconductor detectors. The disadvantages of HPGe detectors are that their lower efficiency than that of NaI(TI) and the necessity to cool it down to the liquid nitrogen temperature.

The low level gamma spectrometry is a very useful technique that needs a great care and highly skills persons to avoid the different source of interferences. The detector size, the source sample geometry, the contents of gamma-ray emitters in the source sample, and the number and the energy of the gamma-rays line emitted most probably produce a complex gamma-ray spectrum in which gamma-rays peaks should be carefully identified and analyzed. For simplicity, gamma-ray spectrum for source emits only one gamma-ray, the spectrum will certainly show following:

The photo-peak (full absorption energy peak) at energy E, which is the characteristics energy for the measured radio-nuclides, The Compton edge; the scattered (compton) electrons have an energy from zero up to a maximum energy $T_{max,}$ which is corresponding to the energy of the compton edge and is given by,

$$T_{max} = E - \frac{E}{1 + \left(\frac{2E}{m_0 c^2} \right)}$$

Backscatter peak; sometime Compton interaction occurs very close to the surface of the detector where is a possibility of electron to escape outside the detector sensitive volume and only the scattered photon will be detected. The minimum energy of the scatted photon corresponds to the energy of the backscattered peak and is given by,

$$E_{\min} = \frac{E}{1+\left(\frac{2E}{m_0c^2}\right)}$$

- The single-escape peak with energy = E – 0.511 MeV, which corresponds to escape of one annihilation photon,
- The double- escape peak with energy = E – 1.02 MeV, which corresponds to escape of both annihilation photons.
- Peak at 511 keV, appears when the pair production reaction occurs near the detector surface and corresponds to the possibility that only one of the annihilation photon detects in the counter.

For high quality gamma-rays spectrometry using HPGe detectors, there are different aspects that should be very carefully considered such as:

Detector Type and Specifications

- Selecting suitable detector type depends on detector counting efficiency, energy resolution and analysis cost, and the right choose of detector size, shape and peak to Compton ratio.
- Detector shields and background reduction,
- Cosmic-ray contribution to background,
- Radon gas concentration in the surrounding counting room environment,
- Electronic setting such as amplifier time constant and plus shaping, pole-zero (P/Z) cancellation and pulse pileup.
- Energy calibration and efficiency calibration of the gamma spectrometer for certain source sample geometry,
- Spectrum analysis, and activity and error calculation

Energy Calibration of Gamma-ray Spectrometer

It is a relationship between the channel number of the multichannel analyzer and the gamma-ray energy, which is essential for gamma-ray spectrum analysis. Mixed standard gamma-ray source, of known radio-nuclides and defined gamma-ray energy lines with at least three different gamma-ray energies to cover the energy range of interest, is used for energy calibration.

Photo-peak efficiency is a function of gamma-ray energy, the source-sample geometry and source sample- detector arrangement. Any slight change in those factors affects the counting efficiency and then affects the accuracy of the calculated activity concentration of the different radio-nuclides in the measured sample. Experimentally, peak efficiency is a function of gamma-ray energy. Efficiency calibration of the spectrometer could be carried out using different methods; the simplest method is using a standard source of the same radio-nuclides of interest such as Cs-134/Cs-137 standard source and the same sample geometry and sample-detector arrangement. The counting efficiency and the activity concentration (A_i) of a radionuclide (J) are given by;

$$\varepsilon_{ij} = \frac{C/t}{A} 100$$

$$A_i = \frac{C_j}{t\,\varepsilon_j\, w}$$

where:

εij = peak efficiency,

C/t = count rate, cps,

A_i = the activity concentration in Bq/kg or Bq/l,

Cij = total count of the photo-peak of radionuclide i at energy j (keV),

t = counting time, s,

w = weighing factor (equivalent to1/sample weight in kg or 1/sample volume in liter),

A = activity concentration of the standard sample, Bq.

Another efficiency calibration method is performed using a standard mixed radio-nuclides source.

Another efficiency calibration method (in the energy range from 186 KeV to 2450 KeV) is performed through two stages. In the first stage the relative efficiency curve of the HpGe detector draws using a 226Ra point source where the most intensive gamma-rays of 226Ra in equilibrium with its daughters have been used. The relative intensities (the intensity of line x divided by the intensity of 609 keV gamma energy line) of the photo-peaks corresponding to these gamma-ray lines have been measured by the detector and calculated. The photo-peak relative efficiency is given by dividing the relative intensity of the photo-peak with energy (E) on the reference relative intensity of the same photo-peak, *i.e.*,

$$\varepsilon(E) = I_M(E)/I_R(E)$$

where:

ε(E) = the relative efficiency at energy (E),

I_M = the relative intensity measured by the detector for the photo-peak with energy (E),

I_R = reference relative intensity of the same photo-peak

The relative efficiency curve of the detector was made of 17 different energy values covering the energy range from 186 KeV to 2450 KeV. The efficiency curve is plotted, see figure 25. The relative efficiency of the detector corresponding to any photo-peak energy can be then obtained using this averaged curve.

In the second stage the average relative efficiency curve of the detector was normalized to an absolute efficiency figure. The normalization has been done using standard solutions of potassium chloride. Naturally potassium, containing 0.0118 % of potassium-40, has specific activity of about 850 pCi/g (31.45 Bq/g). Pure KCl is an excellent low-level reference source in many respects: environmental low - level specific activity, wide availability at high purity and relatively simple branching decay. The normalizing factor for any radionuclide was calculated relatively to the potassium chloride solution normalizing factor using the following equation;

$$N.F(Y) = \frac{R.E(1460keV) \times P.D(K)}{R.E(Y) \times P.D(Y)} \times N.F(K)$$

where:

N.F. (Y) = normalizing factor for (y) radionuclide (Bq.kg-1/cps)

R.E.(1460 KeV) = relative efficiency of 40K,

R.E. (y) = relative efficiency of radionuclide (y)

P.D. (K) = percentage of photon per disintegration of 40 K,

P.D. (y) = percentage of photon per disintegration of (y) radionuclide,

N.F. (K) = normalizing factor of 40K = cps of KCl solution divided by K-40 concentration in Bq/l or Bq/kg.

Counting efficiency varies with variation of the sample density, chemical composition, and geometry and sample-detector arrangement. The effects of sample density and chemical composition on the counting efficiency are energy dependent. To avoid the possible effects of a sample density and composition on the counting efficiency, it is preferred that the counted samples be similar or close to that of the efficiency calibration samples.

Beta-Particles Spectrometry

When radio-nuclides emit no or very low intensity gamma or X-rays, the beta or alpha particles measurement is a necessity. Such beta particles emitter radio-nuclides (maximum beta particle energy in MeV) that are frequently assayed include ^{3}H (0.0186), ^{14}C (0.156), ^{32}P (1.71), ^{35}S (0.167), ^{45}C (0.257), ^{85}Kr (0.84), ^{89}Sr (1.49), ^{90}Sr (0.546), and ^{90}Y (2.28). Because of the continuous nature of a beta particle energy spectrum, spectrometric identification and quantification of a sample containing several beta particle emitters are generally difficult. Thus, radiochemical separation of the different beta particle emitters in environmental samples is essential. Except the conversion electron emitters, they have a sharp energy spectrum. The most often used counting systems for beta particles emitters assay are the gas filled counter operating in either the proportional or Geiger-Muller voltage range and the liquid scintillation counters (LSC). Often the counting systems that are based on gas filled detectors can be more sensitive because of its lower background than that of the LSC. The counting efficiency of the beta particles counting system is given by;

$$\eta = \frac{n}{f_a A}$$

where;

n = the count rate, cps,

f_a = absorption correction factor including the correction factors due to counting geometry factor, source self-absorption and source-mount backscattering,

A = the activity of radionuclide of interest.

It is very essential in the sample source preparation to have the source sample as a thin layer and to be prepared in reproducible technique to minimize the uncertainty due to the change of the absorption correction factor of the standard samples used for efficiency calibration and the actual samples.

Semiconductor detectors, ion implanted detectors, can be used for beta particle spectrometry specially the spectrometry measurement of conversion electron (EC) emitters such as the assay of ^{137}Cs by measuring the 624 keV EC of its ^{137m}Ba daughter.

Alpha-Particles Spectrometry

Many natural occurring radio-nuclides, particularly members of the uranium, thorium and actinium decay series, and the transuranic elements are mainly alpha emitters. Alpha particles spectrometry is a very sensitive technique with a very low counting background. There is a clear different between the alpha particle and other kind of radiation where it has high linear energy transfer and high energy range typically between 4 and 9 MeV. The counting background of the spectrometers may be either due to the alpha particle active materials in their construction material or due to contamination. The background count rates, counting efficiencies and MDA values for different alpha particles counting techniques are given in Table 6.2.

Table 6.2: Background Count Rate

	Background (cpm)	Counting Efficiency (%)	MDA ($Bq.l^{-1}$) in 100 min.
Gas flow proportional counting	0.05 – 0.5	35	2.1
LSC with PSA	0.1 – 1.0	95	0.5
Alpha Spectrometry	0.003 – 0.01	25	0.4

To increase the sensitivity of the alpha particle counting system either to use a detectors having large active areas, or to concentrate the alpha particle active material in a source sample. The thickness of the alpha source affects the quality of alpha particle counting and spectrometry methods. Alpha gross counting can be performed using gas filled detector or LSC. For high alpha particle spectrometry a very thin source sample preparation is essential to reduce the self-absorption of the alpha particles in the source itself and the source sample counts in a vacuum chamber to reduce the absorption of alpha particle in the air between the sample and the detector.

Counting efficiency and energy resolution, checked using a mixed standard alpha emitters electroplated source that contains ^{239}Pu, ^{241}Am and ^{244}Cm or ^{234}U, ^{238}U, ^{239}Pu and ^{241}Am. The counting efficiency does not change with alpha particles energies, *i.e.* it is constant over alpha particles energy range (4-9 MeV). The energy resolution is given as the full width at the half maximum (FWHM) of the ^{241}Am alpha peak at 5.49 MeV. Counting efficiency and energy resolution are affected by the detector active area and source to detector distance. Some alpha particle spectrometer is based on Frisch grid ionization chamber, which usually has a larger active area and a lower energy resolution than that of PIPS detector.

The main advantage of alpha spectrometry is it shih sensitivity due to the high-yield alpha decay process, low background and elimination of other possible interferences by chemical separation. Also it is applicable to a wide range of radio-nuclides and environmental sample. The main disadvantage is the lengthy chemical separation and source preparation procedure.

Radiochemical Analysis

Radiochemical analysis becomes essential when the activity concentration of the radio-nuclides is close to the detection limit of the direct measuring technique or when the levels of interferences in the radionuclide assay process cannot be tolerated. The interference may be due to matrix interference in which the sample matrix absorbs the emitted particles and/or spectral interferences where the spectra of other radio-nuclides prevent the accurate measurement of the radio-nuclides of interest. Radiochemical analysis is essential for a quantitative alpha and beta particles spectrometry of the different natural (such as U isotopes, Th isotopes, ^{210}Pb, ^{210}Po, ^{226}Ra, and ^{228}Ra) and artificial radio-nuclides (such as $^{239+240}$Pu, ^{238}Pu, ^{241}Am and ^{244}Cm) in the different environmental matrixes. Also, improving sensitivity in radiochemical separations usually entails a more through elimination of interfering components.

In general, it is necessary to add isotopic or non-isotopic carriers before any radiochemical analysis is commenced. Isotopic carriers serve two possible functions; to provide additional mass of the element which aids the separation process and to provide a means of measuring the chemical yield. Non-isotopic carriers serve only to aid the separation of the desired nuclides. The sample preparation step includes drying, ashing and scavenging while sample solubilization and equilibrium step includes dissolution and leaching. Radiochemical concentration and separation step includes co-precipitation, ion exchange and solvent extraction. After separation steps, the radio-nuclides of interest exists in a pure portion and finally used for counting source preparation by auto-plating (as for polonium isotopes), electroplating or precipitation (as for U isotopes, Pu isotopes and Th isotopes).

Determination of Uranium Isotopes

The ashed sample is spiked with uranium tracer (^{232}U) for chemical yield and activity calculation. The ashed sample is dissolved using HNO_3, HCl and HF acids. Uranium in the dissolved sample solution is extracted from most of the matrix elements with TOPO (Trioctylphosphine oxide) and back-extracted with 1M NH_4F/0.1 M HCl solution. The uranium fraction is purified by co-precipitation with LaF_3 and anion exchange. Finally, the pure uranium fraction is electro-deposited on a stainless steel disk from HCl/oxalate solution and measured by mean of alpha spectrometry.

Glossary

- **Anti-proton:** The anti-particle of the proton. Identical to the proton except that its charge is opposite in sign.
- **Gamma-ray:** Energetic electromagnetic radiation emitted by radioactive nuclei, produced by particle accelerators and present also in cosmic rays.
- **Mesons and baryons:** Sub-atomic particles, usually with very short lifetimes, believed to be composed of quarks in various combinations.
- **Omega minus particle:** A short lived baryon believed to be made up of three quarks called strange quarks.
- **Photomultiplier tube:** An electronic tube, sensitive to very small amounts of light. The tube converts a light signal into an electrical signal of useful size.
- **Positron:** The anti-particle of the electron. Identical to the electron except that its charge is opposite in sign.
- **Quarks:** Believed to be the most fundamental units of protons and neutrons.
- **Bending magnets:** Wire chambers are used along these beam lines to

actually track individual particles as they move along the beam line. The chambers are similar in a general way to the Geiger counter since they are gas counters. Instead of one wire, the chambers have many parallel wires spaced at distances of a few millimeters. The position of charged particles passing through the chamber can be measured with uncertainties even less than the wire spacing, using fast timing circuits. These chamber measurements facilitate identification of the particle and the measurement of its momentum.

- **Large layered detectors:** The ultimate in particle detectors are probably those being used and constructed at large national and international laboratories such as Fermilab (Fermi National Accelerator Laboratory) in Batavia, Illinois, and CERN in Geneva, Switzerland.

References

Das, Ashok, and Thomas Ferbel. Introduction to Nuclear and Particle Physics. River Edge, NJ: World Scientific, 2004.

Debertin, Klaus; Helmer, Richard G. Gamma- and X-ray Spectrometry with semiconductor detectors, 1st Ed.; Elsevier science Publisher: Amsterdam, 1988; 338 pp.

Folan, Lorcan M. Modern Physics and Technology for Undergraduates. River Edge, NJ: World Scientific, 2003.

Gilmore, Gordon; Hemingway, John Practical Gamma-Ray Spectrometry, 1 st Ed.; John Willy & Sons: New York, 1995; 290 pp.

Green, Dan. The Physics of Particle Detectors. Cambridge, UK, and New York: Cambridge University Press, 2000.

Knoll, Glenn F. Radiation Detection and Measurement, 2 nd.; John Wiley & Sons: New York, 19 ; 735 pp.

L'Annunziata, Michael F. Handbook of radioactivity analysis, 2nd Ed.; Academic Press: Amsterdam, 2004;1062 pp.

Measurement of radio-nuclides in food and the environment, A guidebook, Technical reports series No. 295; International Atomic Energy Agency (IAEA); Vienna, 1989, 169 pp.

Miller, Dudley G. Radioactivity and Radiation Detection, 1st Ed.; Gordon and Breach Science Publishers: New York, 1972; 100 pp.

National council on radiation protection and Measurement (NCRP), 1985. A handbook of radioactivity measurement procedures. NCRP report No. 58, 590 pp.

Serway, Raymond, Jerry S. Faughn, and Clement J. Moses. College Physics. 6th ed. Pacific Grove, CA: Brooks/Cole, 2002.

Taylor, John R., Chris D. Zafirates, and Michael A. Dubson. Modern Physics for Scientists and Engineers. Pearson Prentice Hall, 2004.

Theodorsson, P. Measurement of Weak Radioactivity, 1st Ed.; World Scientific: Singapore, 1996; 333 pp.

Tsoulfanidis, Nicholas. Measurement and Detection of Radiation, 1st Ed.; Hemisphere Publishing Corporation: New York, 1983; 422 pp.

Valkoic Valdo. Chapter 5. Radioactivity in the environment, 1 st Ed.; Elsevier science Publisher: Amsterdam, 2000; 117-207.

Young, Hugh D. Sears and Zemansky's University Physics. San Francisco, CA: Pearson Addison Wesley, 2004.

7

Radiotherapy and Imaging

7.1 X-ray Production, Emission & Interactions with Matter

7.1.1 X-Ray Production

X-rays are simply high-energy photons produced by bombarding a metallic anode with electrons. The production and scattering of X-rays gives additional support to the quantum view of electromagnetic radiation. X-rays are produced when rapidly moving electrons that have been accelerated through a potential difference of the order of 10^3 to 10^6 V strike a metal target. They were first produced by Röntgen in 1895, using an apparatus shown in Figure 7.1.

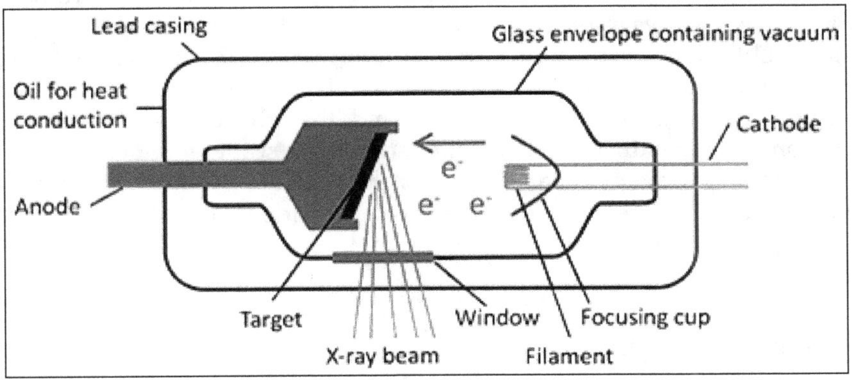

Figure 7.1: Röntgen Apparatus.

Electrons are "boiled off" from the heated cathode by thermionic emission and are accelerated toward the anode (the target) by a large potential difference V. Most of the air is pumped out of the bulb so that electrons can travel from cathode to anode with only a small probability of collisions with air molecules. The residual gas pressure is of the order of 0.01 Pa (10^{-7}atm). When V is a few thousand volts or more, a very penetrating radiation is emitted from the anode surface.

Because of their origin, it is clear that X-rays are electromagnetic waves; like light, they are governed by quantum relations in their interaction with matter. Thus we can talk about X-ray photons or quanta, and the energy of an X-ray photon is related to its frequency and wavelength in the same way as for photons of light, E = hf = hc/. Typical X-ray wavelengths are 0.001 to 1 nm (10^{-12} to 10^{-9} m). X-ray wavelengths can be measured quite precisely by crystal diffraction techniques. X-ray emission is the inverse of the photoelectric effect. In photoelectric emission the energy of a photon is transformed into kinetic energy of an electron; in X-ray production the kinetic energy of an electron is transformed into energy of a photon. The energy relation is exactly the same in both cases. In X-ray production we can neglect the work function of the target because it is ordinarily very small in comparison to the other energies. Two distinct processes are involved in X-ray emission. Some electrons are stopped by the target (anode), and all their kinetic energy is converted directly to a continuous spectrum of photons, including X-rays. Other electrons transfer their energy partly or completely, to individual atoms in the target. These atoms are left in excited states, and when they decay back to the ground state, they emit X-ray photons with energies characteristic of the element in the target. The atomic energy levels associated with excitation by X-rays are rather different in character from those associated with visible spectra. They are associated with vacancies in the inner electron configurations of complex atoms. The energy levels can be hundreds or thousands of electron volts above the ground state, rather than a few electron volts as is typical with optical spectra.

A phenomenon called Compton scattering, first observed in 1924 by A. H. Compton, provides additional direct confirmation of the quantum nature of electromagnetic radiation. When X-rays impinge on matter, some of the radiation is scattered, just as visible light falling on a rough surface undergoes diffuse reflection. Compton discovered that some of the scattered radiation has smaller frequency (longer wavelength) than the incident radiation and that the change in wavelength depends on the angle through which the radiation is scattered. Specifically, if the scattered radiation emerges at an angle Φ with respect to the incident direction, as given and if λ and λ′ are the wavelengths of the incident and scattered radiation, respectively, we find that

$$\lambda - \lambda' = h/mc(1\text{-}\cos\Phi)$$

7.1.2 X-Ray Interaction with Matter

There are two primary means by which we can change the X-ray beam produced by the tube:

- Altering current (mA)
- Altering voltage (kV)

The current (measured in and often referred to as mA, or milliamperes) across the tube determines how many electrons are released to strike the anode. Increasing the mA will increase the number of electrons that strike the anode, with a consequent linear increase in the number of photons produced by the tube. The voltage across the X-ray tube (measured in and often referred to as kV, or kilovolts) affects the velocity of the electrons as they strike the anode; this affects the energy of the photons that can be produced by the tube. Additionally, higher velocity electrons will produce *more* photons, something on the order of $(kV)^3$.

There are two contributions to X-ray spectrum produced by the tube:

- Bremsstrahlung radiation
- Characteristic X-rays

Bremsstrahlung radiation is a continuum with maximal energy at a keV = tube kV. The most photons produced occurs at ~0 keV and almost no photons are produced at keV = kV. Characteristic X-rays are due to electron ejection events; the hole is filled by another electron, which emits an X-ray at a specific frequency. Each element will have one or a couple relevant characteristic X-rays, which are produced at a higher rate compared to the Bremsstrahlung at any frequency (but with a very narrow peak). Due to tube filtration, very low energy photons are removed and do not reach the patient. It is noted that *average energy* of the beam is *much less* than the peak energy; which says 1/3 of the maximum energy.

As observed, there are two ways in which diagnostic X-rays interact with tissue:

- Photoelectric effect; where a photon uses up all of its energy to eject an electron from an atom; while the electron will move around and ionize neighboring atoms, there are no scatter photons.
- Compton scatter; where a photon hits an atom and ionizes an electron but does not use up all of its energy.

The photon then scatters in a different direction with a bit less energy, and the free electron goes about doing damage. Scattered photons can travel back towards the tube, pass through the patient and hit the detector from any odd angle, or scatter again within the patient.

Each time a photon ejects an electron - ionizes an atom - this creates free radicals that can damage DNA and wreak havoc. The energy that a tissue absorbs from photon interactions is referred to as dose.

X-ray Beam Attenuation

As the X-ray beam passes through tissue, photons get absorbed so there is less energy; this is known as attenuation. It turns out that higher energy photons travel through tissue more easily than low-energy photons (*i.e.* the higher energy photons are less likely to interact with matter). Much of this effect is related to the photoelectric effect; the probability of photoelectric absorption is approximately proportional to $(Z/E)^3$, where Z is the atomic number of the tissue atom and E is the photon energy. As E gets larger, the likelihood of interaction drops rapidly.

Compton scattering is about constant for different energies although it slowly decreases at higher energies.

Beam Hardening

As mentioned, low energy X-rays are stopped much more quickly than the high-energy photons. Thus, what comes out of the tissue is mostly the high energy photons referred to as *beam hardening*. The amount of radiation that passes through tissue is given by the Beer-Lambert law, which says that

$$Intensity\ (out) = Intensity(in) * e^{-(\mu * d)}$$

Where

μ = *linear attenuation coefficient*

d = distance the photon has to travel through the tissue

Energy Deposition

As the beam is attenuated - and thus, deposits energy (dose) in the patient - the amount of energy left is less, and the average energy is higher so fewer interactions occur. Both of these effects result in decreasing energy deposition from the skin to the back of the patient. In other words, much of the dose is deposited near the skin. This is why the deterministic effects we are worried about relate to skin injuries. Larger patients, who require higher doses to penetrate through the larger body parts, are at higher risk of skin injuries.

Attenuation and Dose

As you can see from the Beer-Lambert equation, the amount of photons passing through an object drops off rapidly. This is illustrated in the middle panel of the simulation. The thicker the patient or body part, the fewer photons escape to the other side. We need a certain number of photons to produce an image. Back in the days of film, you needed a certain exposure to develop a particular type of film.

Noise

As we realize that each photon being produced is a random event, so the possibility of a photon interacting with tissue and scattering is random. Thus, our image is actually made up of millions of random events. For having enough photons, it turns that in X-rays,

$$SNR \propto 1/\sqrt{N}$$

Where, SNR is the signal-to-noise ratio and N is the number of photons.

kV and Dose

As discussed, high-energy photons pass through tissue with less attenuation. While these photons create more damage when they do hit electrons in the tissue, it turns out that this is a relatively small effect. Thus, overall changing an X-ray beam to a higher energy will deposit less dose to the patient. In addition, because more photons are passing through, there are more photons to make our image - so it will be less noisy. Also, remember that when we crank up the kV, we get more photons from the tube itself (photons $\propto kV^3$).

mA and Dose

Radiography and fluoroscopy systems have so-called photo-timers that shut off the beam after they get enough photons for the desired noise settings. Thus, increasing mA means that the image will take less time (more photons per second -> fewer seconds). Note that X-ray tubes have a maximum mA because they can overheat. This is relevant in long angiography runs or in magnification mammography.

Patient Thickness and Dose

As seen, far fewer photons will make it through a thicker patient - exponentially so. Therefore, we have to send more photons through in the first place. More photons means more likelihood to hit the tissue and thus more dose. Also, larger patients have more tissue for scattered photons to hit a second or third time, increasing dose.

Tissue Contrast

If all different types of tissue stopped X-rays in the same way, then we would have no picture - just a gray blob on our screen. The difference in X-ray penetration between different tissues represents the *contrast* in the image, *i.e.* how easily we can tell different objects apart.

Photoelectric Absorption

As seen, probability of photoelectric absorption is proportional to $(Z/E)^3$ however, is the concept of *k-edge*. Albert Einstein won the Nobel Prize for explaining how photoelectric events require a minimum amount of energy; it turns out that this is also the *most likely* energy for those events to occur. Higher energy photons are much less likely to cause photoelectric events. The k-edge represents the energy needed to eject a K-shell electron. Outer shell electrons have absorption edges but these are much too low energy to be relevant.

In soft tissues, the dominant elements (*e.g.* C, H, O, and N) have very low K-edges, in the range of a few keV. While these elements do contribute to the photoelectric effect and attenuate low energy X-rays, there is no relevant k-edge with its substantial change in attenuation. However, the elements iodine and barium have K-edges around 30-40 keV, right in the middle of the X-ray beam spectrum. Thus, soft tissues with even a small amount of iodine will have a much stronger X-ray stopping power than those without.

Absorption due to Compton scattering is nearly independent of the X-ray energy. It does depend on the material (specifically, the electron density) but is still much less than the photoelectric absorption in diagnostic X-ray imaging. Compton scattering is responsible for the right-hand tail in the above graph. Note that it also creates scatter photons that may hit the image detector and contribute to background/noise as they are scattered at different angles, we do not know where they initially came from).

Thus, the main contribution to differences in X-ray attenuation in diagnostic imaging is photoelectric absorption. You can see from this graph that these differences are much more apparent at *low energies*. This is why lowering kV improves image contrast. The presence of materials with high k-edges such as iodine or barium improve the contrast at higher kV, but contrast is still better at lower kV. For example, a beam at 80 kV will have an average X-ray energy near 30 keV - exactly the k-edge of iodine (Figure 7.2).

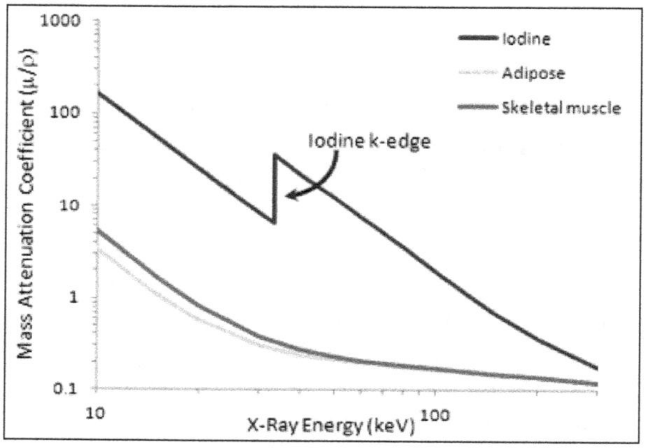

Figure 7.2: Showing Mass Attenuation Coefficients.

Mammography

In mammography, we need to distinguish thin wisps of breast tissue or tiny calcifications from the surrounding fat. Soft tissue and fat attenuation curves are shown above; Calcium has a k-edge of 4 keV. In order to obtain image contrast with these materials, we need to utilize *extremely low energy photons*.

One way of obtaining a very low energy beam is to use a low kV. We also need to make sure that we have no filtration. Many mammography units replace glass with beryllium as the window for the X-ray tube. Beryllium has a very low atomic number (Z=4) so its attenuation is less than glass; additionally, glass (mainly SiO_2) has a k-edge at 1.8 kV and higher atomic numbers (8 and 14), with increased photoelectric attenuation.

In addition to low kV, a more efficient way of producing low-keV X-rays. Traditionally, this was done by using a molybdenum instead of tungsten anode. One component is the Bremsstrahlung radiation, which depends simply on the tube voltage. The other component, which is more efficient, is the characteristic radiation. Characteristic X-rays represent those produced when an electron from the cathode knocks out an electron in an anode atom; another electron from a higher-energy orbital takes its place. This new electron releases an X-ray with the energy of the difference in binding energies of the two orbitals (Figure 7.3).

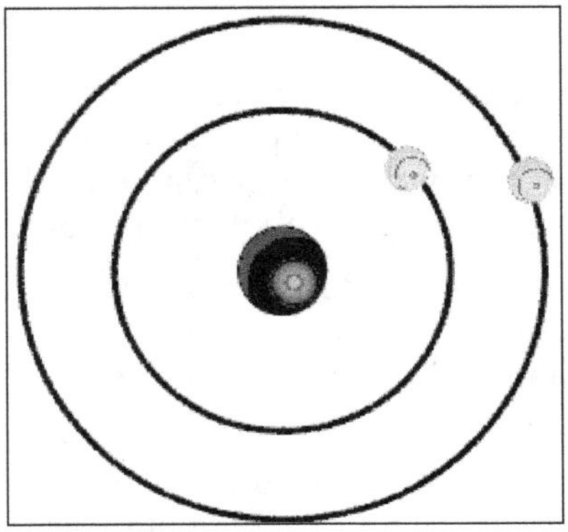

Figure 7.3: Characteristic X-ray Production Process.

Anode Material

For tungsten, the characteristic X-rays are around 60-70 keV, higher than what we want. For molybdenum, the characteristic X-rays are ~17.5 keV and 19.5 keV. These are in exactly the range we want to distinguish fat, soft tissue, and calcium. Now, we will still get the Bremsstrahlung radiation - but those high-energy X-rays will not contribute to contrast. We can use molybdenum as the *filration material* as well as the anode. The k-edge of an element will always be slightly greater than its characteristic X-rays (~ 20 keV).

The k-edge represents the energy needed to knock out a k-orbital electron. The characteristic X-ray represents the energy lost to dislodge a higher-orbital electron with much lower binding energy but also the energy gained by filling the k-shell. Thus, the characteristic X-ray energy is the k-shell binding energy *minus* L-shell binding energy.

An alternative to molybdenum is rhodium, with characteristic X-rays at 20.2 and 22.7 keV and k-edge of 23.3 keV.

Things to Remember

- X-ray tubes produce photons by electron bombardment of a metallic anode. The resulting photon spectrum is a continuum and is determined by
  (a) Characteristic X-rays
  (b) Bremsstrahlung radiation
- The number of photons sent through a patient is determined by the current across the tube (mA) and the imaging time (s), which are sometimes multiplied together as mA*s = mAs. Changing kV changes the number of photons in a non-linear fashion ~ $(kV)^3$.

- X-rays interact with tissue in 2 main ways:

  Photoelectric effect

  Compton scatter
- The photoelectric effect is highly accentuated at the *k-edge* of a material; materials with k-edges in the relevant range include iodine and barium. The k-edges are typically low energies, and the closer your X-ray energy is to the k-edge, the more likely a photoelectric event is to occur.
- Thus, to improve contrast, decrease your X-ray energy (kV). However, this increases dose because more X-rays are absorbed by the patient so you need to send more X-rays through from the tube. This represents the central trade-off in X-ray imaging.
- X-ray penetration is an exponentially decreasing function of patient thickness. Large patients get much larger doses; you can improve this by increasing the X-ray energy but at the cost of contrast.
- Lower energy X-rays are more attenuated for any given distance than high-energy X-rays. Thus, as the beam passes through material, it becomes *harder*, *i.e.* increases in average X-ray energy.

7.2 Radiographic Film, Latent Image, Intensifying Screens, Grids

7.2.1 Radiographic Film

Properties and Storage

Film Speed

- The term "speed" can be applied to films as well as screens Radiography: cassette and intensifying screen.
- A fast film will become blacker for a given exposure than as low film.
- The speed of films is determined by the size and number of halide grains.
- Larger grains give faster speed but reduced resolution.
- Speed is a relative value and cannot be compared across different manufacturers.
- Speed is often used to describe a film-screen combination rather than one or the other in isolation.

Storage

- Radiographic film is a delicate material and requires careful handling.
- Cool conditions (10-18°C).
- Dry conditions (no more than 50% humidity).
- Away from sudden changes in temperature and humidity.

- Protected from light.
- Protected from processing chemicals and their fumes.
- Protected from ionizing radiation.
- Film stored on edge.

Film Identification

- It is essential that radiographs are clearly, correctly and indelibly labelled.
- Information on each film should include
- Owner and patient's name or identification number.
- Date radiograph taken.
- Hospital or practice name.
- Anatomical marking ie left or right.
- Methods of marking film:
- "X-Rite" tape:
- Self-adhesive, lead-impregnated tape on which information is written and then stuck to cassette before exposure.
- Can be difficult to see on finished radiograph, especially if exposure has been high.

Lead Numbers and Letters

- Commonly used for anatomical marking but can also be used for identification.
- Words formed from required combinations of letters and stuck to cassette before exposure.
- Metal holders, into which the letters slide, are also available and can be stamped with practice name.

Light Marker

- Photograph the identification information, written on a special piece of paper, onto a corner of the film which has been shielded from the main exposure by a lead blocker.
- This marking can be done by manual marking of the film in the dark room or if cassette is inserted into machine before film is removed the identification window in the cassette automatically slides back and exposure is made.

7.2.2 Latent Image

It is an invisible image which is produced by an effect of light on matter which can be rendered visible by the subsequent process of photographic development. The photographic latent image may be defined as that radiation-induced change in a grain or crystal that renders the grain readily susceptible to the chemical action of a developer.

Such images are formed due to action of light on silver halide grains in emulsion forms. The basic mechanism by which this happens was first proposed by R W Gurney and N F Mott in 1938. The incoming photon liberates an electron, called a photoelectron, from a silver halide crystal. Photoelectrons migrate to a shallow electron trap site, where the electrons reduce silver ions to form a metallic silver speck. A positive hole must also be generated but it is largely ignored. Subsequent work has slightly modified this picture, so that 'hole' trapping is also considered. Since then, understanding of the mechanism of sensitivity and latent image formation has been greatly improved.

At the time of formation of latent image, a good deal was known about its physical nature. It was known; that, latent image was localized at certain discrete sites on the silver halide grain. If a photographic emulsion is exposed to light, developed briefly, fixed, and then examined under a microscope, it can be seen that development started at only one or few places on crystal. As the small amounts of silver sulfide on surface of grain were necessary for photographic material to have high sensitivity, it seems likely that spots at which the latent image was localized were local concentrations of silver sulfide (Figure 7.4).

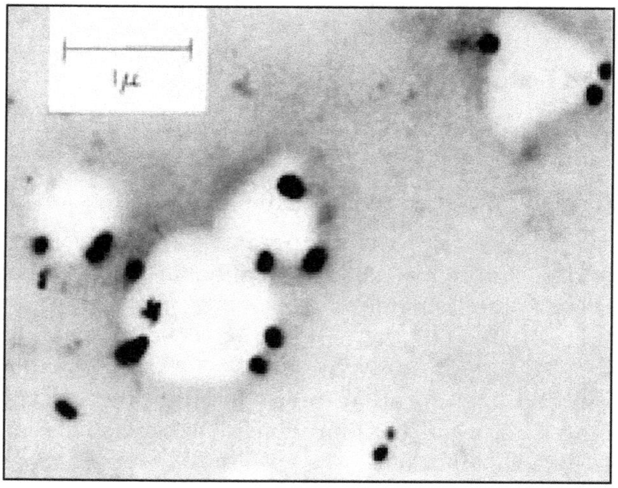

Figure 7.4: Exposed Grains Showing Initiation of Development at Local Sites.

It was further known that material of latent image was in all silver. For one thing, chemical reactions that will oxidize silver will also destroy the latent image. Also, it is common observation that photographic materials given prolonged exposure to light darken spontaneously, without the need for development. This darkening is known as the print-out image. The printout image contains enough material to be identified chemically, and this material is metallic silver. By microscopic examination, the silver of the print-out image is discovered to be localized at certain discrete areas of the grain, just as is the latent image (Figure 7.5).

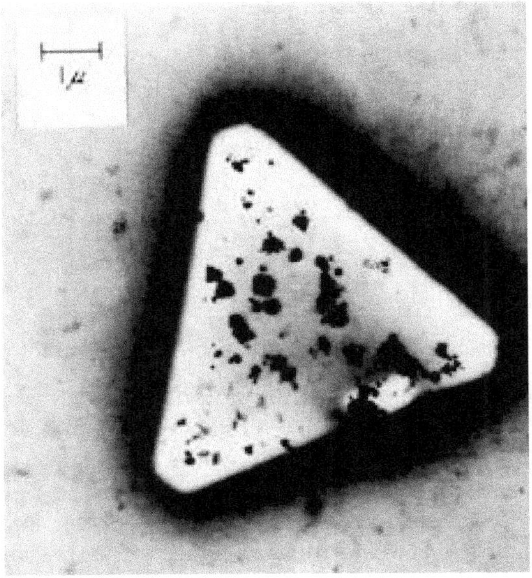

Figure 7.5: Micrograph of Photolytic Silver Produced in a Grain.

Hence, the change that makes an exposed photographic grain capable of being transformed into metallic silver by the mild reducing action of a photographic developer is a concentration of silver atoms--probably only a few--at one or more discrete sites on the grain. Any theory of latent-image formation must account for the way that light photons absorbed at random within the grain can produce these isolated aggregates of silver atoms. Most current theories of latent-image formation are modifications of the mechanism proposed by R. W. Gurney and N. F. Mott in 1938.

For understanding Gurney-Mott theory of latent image, it is necessary to digress and consider the structure of crystals--in particular, the structure of silver bromide crystals. When solid silver bromide is formed, as in the preparation of a photographic emulsion, the silver atoms, each give up one orbital electron to a bromine atom. The silver atoms, lacking one *negative* charge, have an effective positive charge and are known as silver *ions* (Ag^+). The bromine atoms, on the other hand, have gained an electron--a negative charge--and have become bromine ions (Br^-). The "plus" and "minus" signs indicate, respectively, one fewer or one more electron than the number required for electrical neutrality of the atom.

A crystal of silver bromide is a regular cubical array of silver and bromide ions, as shown schematically in the figure below. It should be emphasized that the "magnification" of the figure is very great. An average grain in an industrial X-ray film may be about 0.00004 inch in diameter, yet will contain several billions of ions.

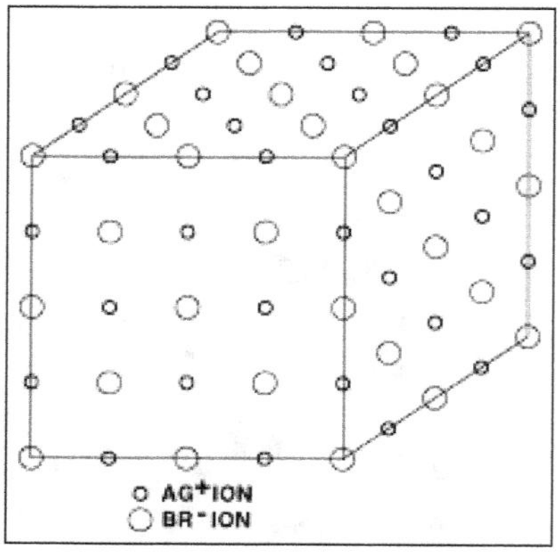

Figure 7.6: Silver Bromide Crystal of Silver (Ag^+) and Bromide (Br^-) Ions.

A crystal of silver bromide in a photographic emulsion, has a number of imperfections as within the crystal, there are silver ions that do not occupy lattice position as shown, but rather are in the spaces between. These are known as interstitial silver ions as in Figure 7.6. The number of the interstitial silver ions is, of course, small compared to the total number of silver ions in the crystal. In addition, there are distortions of the uniform crystal structure. These may be foreign molecules, within or on the crystal, produced by reactions with the components of the gelatin, or distortions or dislocations of the regular array of ions shown in the Figure 7.7. These may be classed together and called latent-images sites.

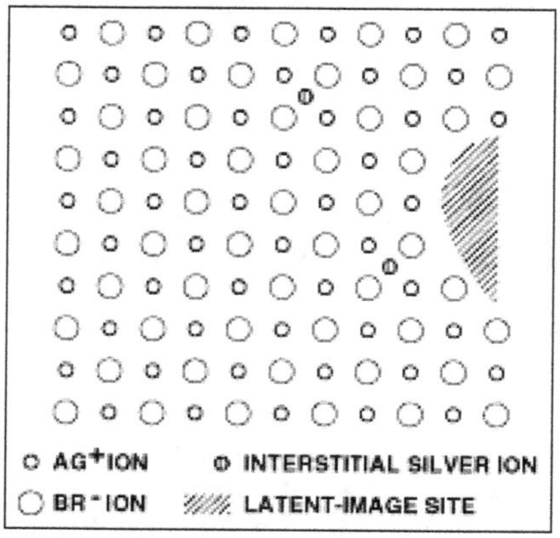

Figure 7.7: View of Ions of Crystal with Two Interstitial Silver Ions.

As per Gurney-Mott theory, when a photon of light of energy greater than a certain minimum value is absorbed in a silver bromide crystal, it releases an electron from a bromide (Br^-) ion. The ion, having lost its excess negative charge, is changed to a bromine atom. The liberated electron is free to wander about the crystal as shown in figure below. As it does, it may encounter a latent image site and be "trapped" there, giving the latent-image site a negative electrical charge. This stage of latent-image formation involves transfer of electrical charges using moving electrons.

Here, negatively charged trap will attract an interstitial silver ion as the silver ion is charged positively. When such an interstitial ion reaches a negatively charged trap, its charge is counteracted, an atom of silver is deposited at the trap, and the trap is reset (D). This second stage of the Gurney-Mott mechanism is termed the ionic condition stage, since electrical charge is transferred through the crystal by the movement of ions--that is, charged atoms (Figure 7.8). The whole cycle can recur several, or many, times at a single trap, each cycle involves absorption of one photon and addition of one silver atom to the aggregate. (E to H).

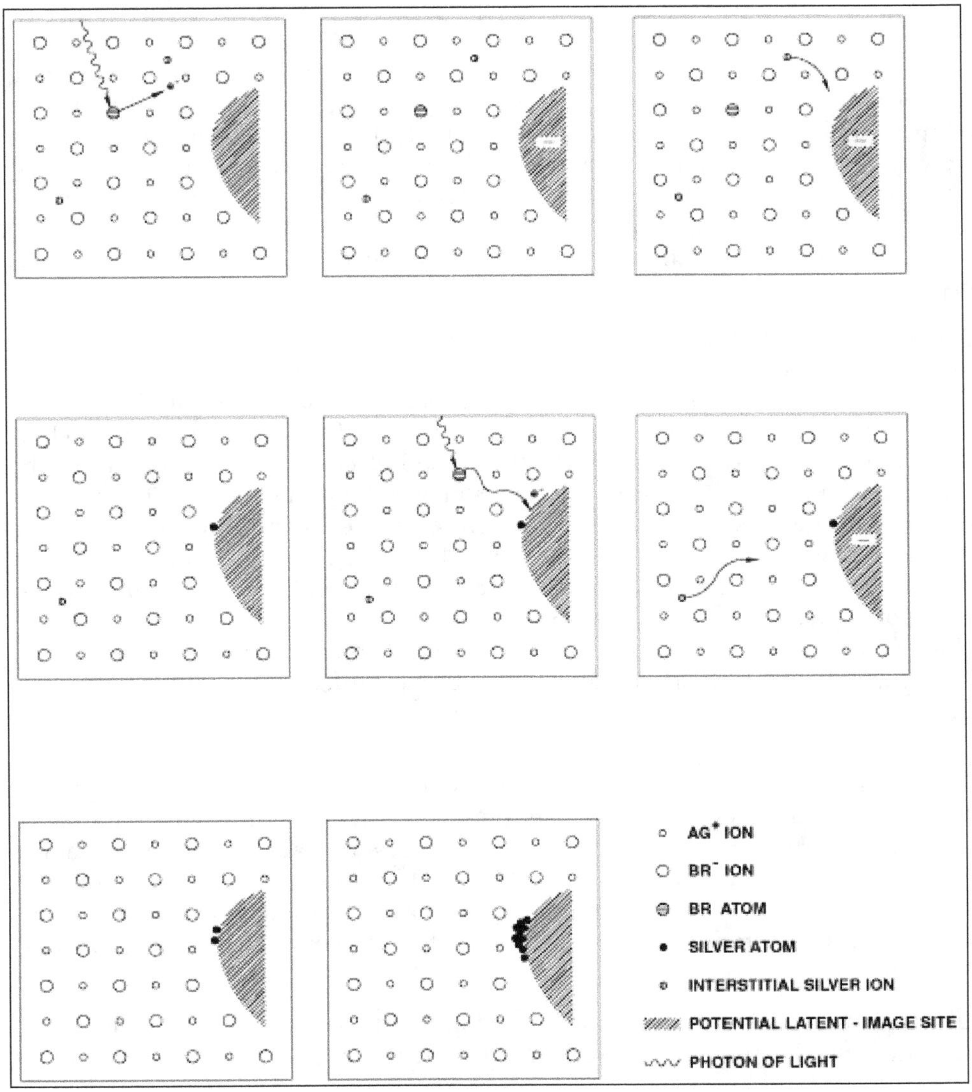

Figure 7.8: Stages in Latent Image Development.

Also, this aggregate of silver atoms is latent image where presence of few atoms at single latent-image site makes the whole grain susceptible to reducing action of developer. In sensitive emulsions, number of silver atoms required may be less than ten.

The mark of success of theory is its ability to provide an understanding of previously inexplicable phenomena. The Gurney-Mott theory and those derived from it have been notably successful in explaining a number of photographic effects.

Low-intensity reciprocity-law failure (left branch of the curve) results from the fact that several atoms of silver are required to produce a stable latent image. A single atom of silver at a latent-image site (D) is relatively unstable, breaking down rather easily into an electron and a positive silver ion. Thus, if there is a long interval between the formation of the first silver atom and the arrival of the second conduction electron (E), the first silver atom may have broken down, with the net result that the energy of the light photon that produced it has been wasted. Therefore, increasing light intensity from very low to higher values increases the efficiency, as shown by the downward trend of the left-hand branch of the curve, as intensity increases.

High-intensity reciprocity-law failure (right branch) is frequently a consequence of the sluggishness of the ionic process in latent-image formation. According to the Gurney-Mott mechanism, a trapped electron must be neutralized by the movement of an interstitial silver ion to that spot (D) before a second electron can be trapped there (E); otherwise, the second electron is repelled and may be trapped elsewhere. Therefore, if electrons arrive at a particular sensitivity center faster than the ions can migrate to the center, some electrons are repelled, and the center does not build up with maximum efficiency. Electrons thus denied access to the same traps may be trapped at others, and the latent image silver therefore tends to be inefficiently divided among several latent-image sites. Thus, the resulting inefficiency in the use of the conduction electrons is responsible for the upward trend of the right-hand branch of the curve.

7.2.3 Intensifying Screens

Intensifying screen is used in the cassette to intensify the effect of the X-ray photon by producing a larger number of light photons. It decreases the mass required to produce a particular density and hence decreases the patient dose significantly. In cassettes, which use double emulsion films, two screens are used, mounted on both sides of the cassette. In mammography, however, a single screen on the back side and a single emulsion film is used. The thickness of an intensifying screen is about 0.4 mm (Figure 7.9).

Intensifying screens are used because the film is much more sensitive to light than to X-radiation; approximately 100 times as much X-radiation would be required to expose a film without using intensifying screens. Unfortunately, intensifying screens introduce blurring into the imaging process and places a limit on the visibility of detail that must be considered when selecting screens for specific clinical applications.

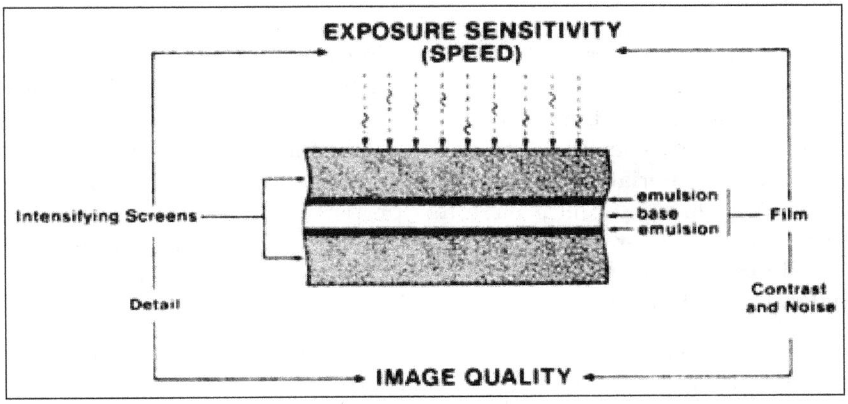

Figure 7.9: Intensifying Screens.

Different types of intensifying screens are available for clinical use. The selection of a screen for a specific procedure is usually based on a compromise between the requirements for image detail and patient exposure. The receptor used for most radiographic procedures contains two intensifying screens mounted on each side of double-emulsion film. Using two screens in this manner increases X-ray absorption and receptor sensitivity with the least amount of image blurring. In some procedures that require high image detail, such as mammography, one intensifying screen is used in conjunction with a single-emulsion film.

Functions

X-Ray Absorption

The first function performed by the intensifying screen is to absorb the X-ray beam (energy) emerging from the patient's body. The ideal intensifying screen would absorb all X-ray energies that enters it; real intensifying screens are generally not thick enough to absorb all of the photons. As we discussed earlier, increasing the thickness of an intensifying screen to increase its absorption capabilities degrades image quality.

In most cases, a significant portion of X-ray energy is not absorbed by the screen material and penetrates the receptor. This is wasted radiation, since it does not contribute to image formation and film exposure. The absorption efficiency is the percentage of incident radiation absorbed by the screen material. An ideal screen would have a 100% absorption efficiency; actual screens generally have absorption efficiencies less than 100%. Absorption efficiency is primarily determined by three factors:

- screen material
- screen thickness
- photon energy spectrum

Light Production

The second function performed by the intensifying screen is to convert a portion of the absorbed X-ray energy into light. This is the fluorescent process. Fluorescence is the property of a material that enables it to absorb radiation energy in one portion of the photon-energy spectrum and emit some of the energy in the form of lower energy photons. Materials that glow, or emit visible light, when exposed to high-photon energy ultraviolet light have this property. The figure below illustrates what happens to the X-ray energy that is absorbed by an intensifying screen. In the intensifying screen, the fluorescent process creates visible light when such material is exposed to high-energy X-ray photons.

The intensifying screen is an energy converter; it converts approximately 5 to 20% of the absorbed X-ray energy into light. This percentage is the conversion efficiency of the screen, and depends on the type of material used in the screen. Although the total energy of the light emitted by a screen is much less than the total X-ray energy the screen receives, the light energy is much more efficient in exposing film because it is "repackaged" into a much larger number of photons. If we assume a 5% energy conversion efficiency, then one 50-keV X-ray photon can produce 1,000 blue-green light photons with an energy of 2.5 eV each.

Exposure Reduction

Since film is more sensitive to light than to X-ray exposure, film can be exposed with much less radiation if an intensifying screen is used. Conventional X-ray film has an X-ray exposure sensitivity in the range of 50 mR to 150 mR if exposed directly by the X-radiation. When the film is combined with intensifying screens, the sensitivity ranges from approximately 0.1 mR to 10 mR, depending on the type of screen and film used.

Receptor Sensitivity

The sensitivity of a receptor, such as an intensifying screen-film combination, is expressed in terms of the exposure required to produce a film density of 1 unit above the base plus fog level. Some manufacturers do not provide sensitivity values for their receptor systems, but most provide speed values such as 100, 200, 400, *etc.* The speed scale compares the relative exposure requirements of different receptor systems. Most speed numbers are referenced to a so-called par speed system that is assigned a speed value of 100. Whereas sensitivity is a precise receptor characteristic that expresses the amount of exposure the receptor requires, speed is a less precise value used to compare film-screen combinations.

There is, however, a general relationship between exposure requirements (sensitivity) and receptor speed values:

Sensitivity (mR) = 128/speed.

For example, a receptor with a true speed value of 100 requires an exposure of 1.28 mR to produce a 1-unit film density. Sensitivity and speed values are inversely related. A more sensitive receptor has a higher speed value than a less sensitive receptor. The range of receptor sensitivity and speed values used in radiography is shown below.

Speed	Sensitivity (mR)
1200	0.1
800	0.16
400	0.32
200	0.64
100	1.28
50	2.56
25	5.0
12	10.0

Most receptors are given a nominal speed rating by the manufacturer. The actual speed varies, especially with the X-ray spectrum (KV) and film processing conditions. The sensitivity (speed) of an intensifying screen-film receptor depends on the type of screen and film used in addition to the conditions under which they are used and the film is processed.

Characteristics

Spectral Characteristics

The other elements in the compound contribute to the fluorescent properties of the material. Each compound produces light of a color (wavelength) that is specific to the particular material. The light from intensifying screens is produced in either the ultraviolet, blue or green portion of the light spectrum, and intensifying screens are sometimes classified as either blue or green emitters. The significance of this is that a screen must be used with a film that has adequate sensitivity to the color of light the screen emits. Some radiographic films are sensitive only to blue light; others (orthochromatic) are also sensitive to green light. If screen and film spectral characteristics are not properly matched, receptor sensitivity is severely reduced.

Thickness

The selection of a screen is generally a compromise between exposure and image quality, as illustrated in Figure 7.10. Thin screens absorb a relatively small fraction of the X-ray photons; thicker screens absorb a greater fraction and thus require less X-radiation to produce the same film exposure. Unfortunately, increasing screen thickness also increases image blur.

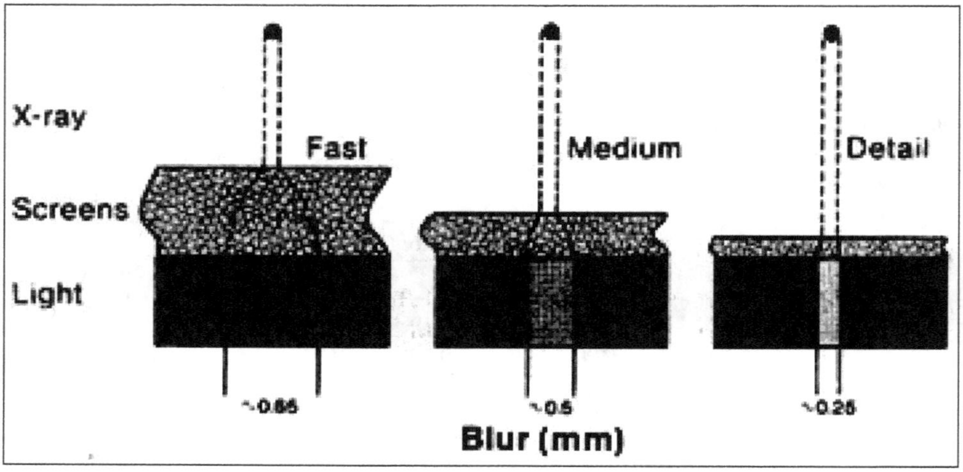

Figure 7.10: Effects of Screen Thickness on Image Blur.

Photon Energy

The sensitivity of intensifying screens varies with X-ray photon energy because sensitivity is directly related to absorption efficiency. Absorption efficiency and screen sensitivity are maximum when the X-ray photon energy is just above the K edge of absorbing material. Each intensifying screen material generally has a different sensitivity-photon energy relationship because the K edge is at different energies. The spectrum of photon energies within an X-ray beam is most directly affected and controlled by the KV; therefore, the sensitivity and speed of a specific intensifying screen is not constant but changes with the KV selected for a specific procedure. Significant exposure errors can occur if technical factors (KV and MAS) are not adjusted to compensate for the variation in screen sensitivity. This often occurs when the same technique charts are used with screens composed of different materials. Also, the KV response characteristics of automatic exposure control (AEC) sensors should be matched to those of the intensifying screens.

Image Blur

The most significant effect of intensifying screens on image quality is that they produce blur. The reason for this was illustrated above. Considering imaging of very small object, like calcification, where X-ray photons passing through the object gets absorbed and produce light along the vertical path extending through the intensifying screen.

Before exiting the screen, the light spreads out of the absorption path, as illustrated. The light image of the object that appears on the surface of the intensifying screen is therefore blurred; the degree of blurring by this process is related to the thickness and light transparency of the intensifying screen. The major issue in selecting intensifying screens for a particular clinical application is arriving at an appropriate compromise between patient exposure and image quality or, more specifically, between receptor sensitivity (speed) and image blurring. Screens that produce maximum visibility of detail generally have a low absorption efficiency (sensitivity) and require a relatively high exposure. On the other hand, screens with a high sensitivity (speed) cannot produce images with high visibility of detail because of the increased blurring (Figure 7.11).

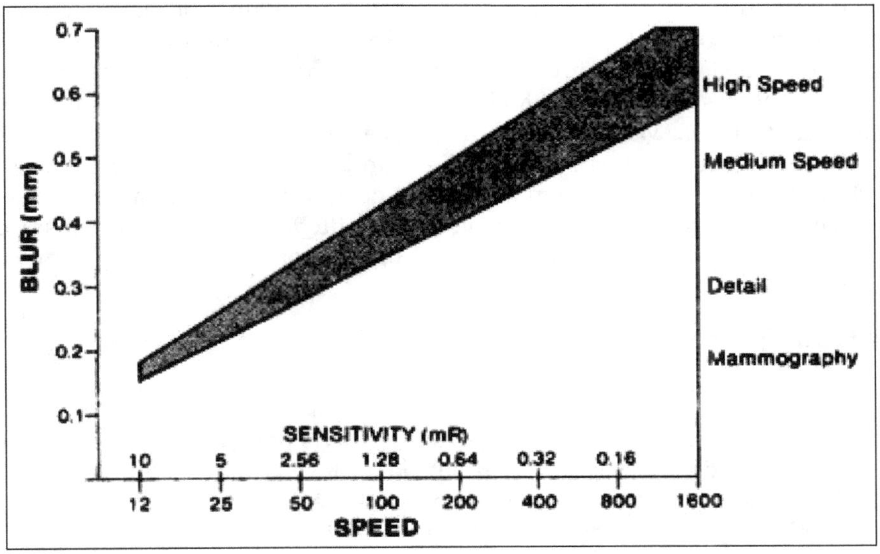

Figure 7.11: Relationship between Image Blur and Sensitivity.

Screen-Film Contact

If the film and intensifying screen surfaces do not make good contact, the light will spread, as shown below, and will produce image blurring. This is an abnormal condition that occurs when a cassette or film changer is defective and does not apply sufficient pressure over the entire film area. Inadequate film-screen contact usually produces blurring in only a portion of the image area (Figure 7.12).

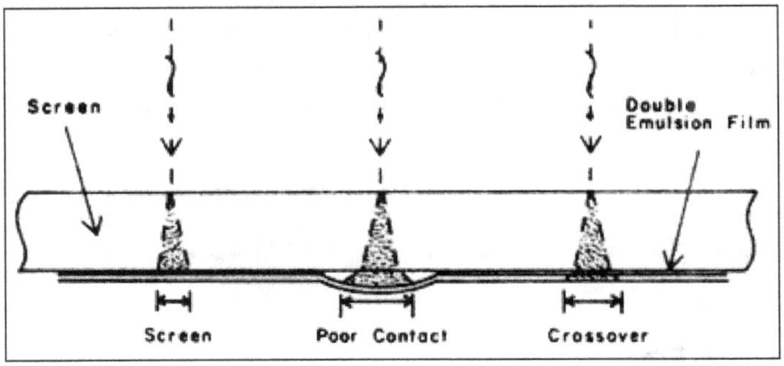

Figure 7.12: Sources of Blur in Screen-Film Receptors.

The conventional test for film-screen contact is to radiograph a wire mesh. Areas within the image where contact is inadequate will appear to have a different density than the other areas. This variation in image density is most readily seen when the film is viewed from a distance of approximately 10 ft and at an angle.

Crossover

If the film emulsion does not completely absorb the light from the intensifying screen, the unabsorbed light from one side can pass through the film base and expose the emulsion on the other side. This is commonly referred to as crossover. As the light passes through the film base, it can spread and introduce image blur, as illustrated above. Many modern film-screen receptor systems are designed to minimize crossover blurring. Crossover can be decreased by placing a light-absorbing layer between the film emulsion and film base, using a base material that selectively absorbs the light wavelengths emitted by the intensifying screens, and designing the film emulsion to increase light absorption. It is not a significant problem with modern radiographic systems.

7.2.4 Grid

Introduction

A grid is a device that, when introduced between the patient and the film cassette (or image intensifier), reduces the amount of scattered radiation that can reach the film cassette. The grid was invented by Gustave Bucky in 1913 and is still the most effective device for reducing scattered radiation. It is usually made of a thin plate, in which very thin strips of high-attenuating material (commonly lead) is placed in a linear, parallel pattern. The spaces between the strips are filled with low-attenuating material. In older grids this material can be paper or aluminium, but in modern grids it is usually made of carbon fibre. The lead strips are always higher than the distance between them. The grid ratio is defined as the ratio of the height of the lead strips to the distance between them, and is, accordingly, always greater than unity.

Scattered radiation is the biggest factor that contributes to poor diagnostic quality of radiographs. Its effect produces a general radiographic fog on the film which reduces the contrast. The best-known way to effectively remove the greater part of radiation scatter is by the use of an X-ray anti-scatter grid. Radiation which does not travel in the same direction as the primary beam is absorbed by the lead strips of the grid.

Types

Grids are divided into four types based on their applications like:

- Parallel types
- Focused types

- Crisscross types
- Tapered types

A single grid may be of more than one type, but this is rare. The majority of grids used in general radiography are focused grids, focused for 40″, 40-72″, or 72″ distance from the X-ray tube, but other applications require different grids.

Grid performance is calculated by comparing the percentage of radiation that gets through the grid (Total Transmission), the percentage of scattered radiation that gets through the grid (Scatter transmission) and the percentage of radiation that has not been scattered that gets through the grid (Primary transmission). If the grid is perfectly focused, then estimate of these can be calculated from specifications of the grid (Figure 7.13).

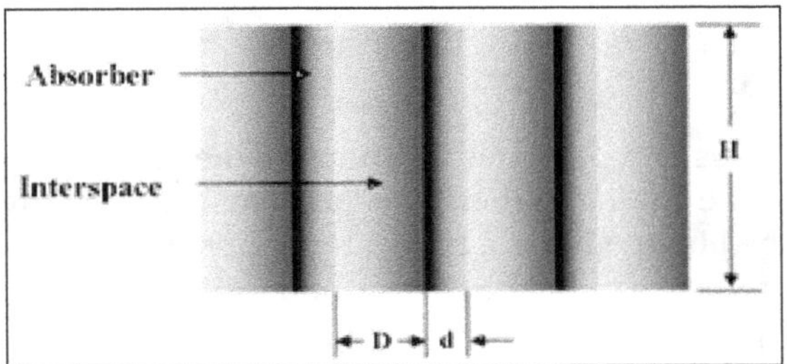

Figure 7.13: Performance Factor of Grid.

Parallel Type

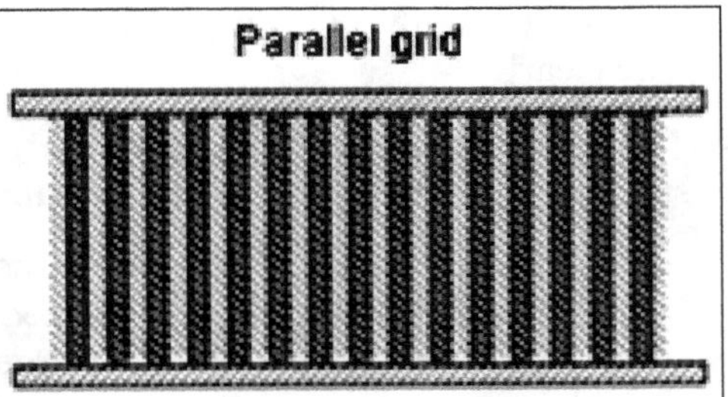

Figure 7.14: Parallel Grid.

It is a grid where absorbing strips are parallel to each other in their longitudinal axis. Most linear grids are also focused, *i.e.* their strips are slightly tilted, converging at a line in space (the convergent line). A non-focused linear grid will have strips that are parallel when viewed in cross-section; this is called a parallel grid. Many X-ray tables are equipped with linear, focused grids, and the strips in these grids are parallel with the long axis of the table, allowing the X-ray tube to be tilted in this direction without changing the effectiveness of the grid (Figure 7.14).

Focused Grid

It is a grid where absorbing strips are slightly angled towards the focal spot. The grid can be used only at specified focal distance or the grid will absorb the primary radiation and parts of the film which are barely exposed. Focused grids may be linear or crossed.

Criss Cross Type

A grid consisting of two superimposed parallel grids; having similar focusing distance. Such grids are very efficient in removing scattered radiation but must be arranged at exactly the right angle to the beam. The use of such girds is therefore limited.

Tapered Type

It is a grid where the surface is tapered to peak at center of grid that functions similar to focused grid. All of the strips are parallel to each other, and the tapered surface is toward the focal spot type (Figure 7.15).

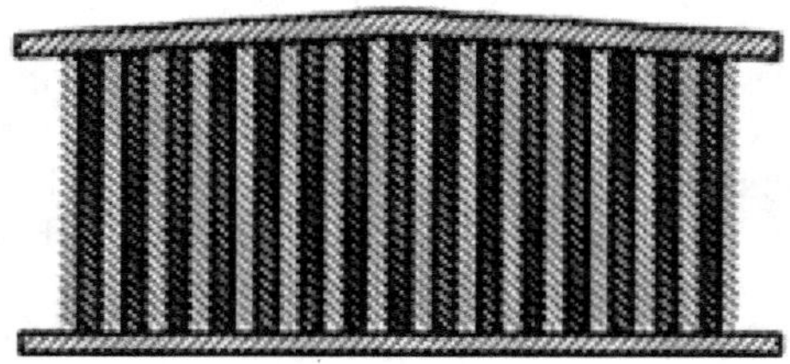

Figure 7.15: Tapered Grid.

Mechanism of Grid

Whenever X-rays pass through matter, some of the X-ray photons will interact with the atoms of the matter. When you take an X-ray, many incoming X-ray photons are absorbed by the bone, a high attenuation material, while other tissues absorb less. This produces an image of dark and light on the X-ray film. The effect is like holding up paper to a light to see the writing on the other side. However, absorption is not the end. As part of the absorption, a second photon with less energy will be produced, flying off in a different direction than the photon that was absorbed. These "scattered" photons will create a random grayness on the image, reducing the contrast between body tissues and making it hard to read the image clearly.

In order to prevent the shadows cast on film by grid from interfering with visualization of diagnostic detail, certain principles must be followed:

- For one, the lead should be as thin as possible to be consistent with adequate absorption of scattered radiation. The thinner the lead, the narrower the shadow it will produce on the film and the less visible it will be to the eye.
- Also, the thinner it is the less absorption of primary radiation will be in the grid. However, it must be noted that adequate absorption of scattered radiation is the function of the grid and lead must be thick enough to provide this function.
- Another factor is the relative fineness of the grid. This quality is represented by the number of lines per inch. In general, the greater the number of lines per inch, the less visible will the individual lines be, but this is subject to certain practical considerations which modify it in actual use.

The selection of a grid to be used for a particular radiograph will be primarily dependent on the following considerations:

- Relative quantity of scattered radiation produced by subject being radiographed.
- Kilo voltage technique used.
- Capacity of X-ray generator.

X-ray grids are commercially available with either focused or parallel lead strips, and these two types are produced in either linear or crossed grid configuration. X-ray Grid Selection Based on Clean-up Requirements are:

Cleanup	Ratio/ Type	Positioning Latitude	Recommended Up To	Remarks
Superlative	8:1 criss-cross	Distance fair; centering and leveling-slight	120 KVP	Not recommended for tilted tube technique
Excellent	12:1 linear	Very slight	110 KVP (Suitable for highr KV)	Extra care required for proper alignment; usually used in fixed mount
Excellent	6:1 criss-cross	Good	100 KVP	Tube tilt limited to five degrees
Good	8:1 linear	Distance fair; centering and leveling-slight	100 KVP	For general stationary grid use
Good (CR)	4:1,6:1	Good	120 KVP	CR grid
Moderate	6:1 linear	Good	80 KVP	Least expensive of stationary grids

The focused grid has its leads angled progressively in such a way that lines drawn through each lead and continued out of the grid will intersect at a point known as the grid focus. When strips are not progressively angulated but are perpendicular to the surface of the grid, the grid is termed as parallel (Figure 7.16).

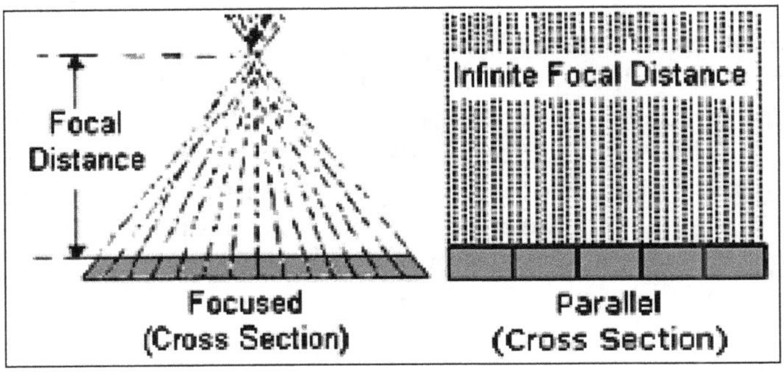

Figure 7.16: Parallel and Focused Grids.

Both the focused or parallel grids may be made in either the linear or crossed grid type. The linear grid is made with the lengths of all its leads in the same direction. The crossed grid is usually two linear grids, one on top of the other, with the leads of the top grid crossing those of the lower grid as shown in Figure 7.17.

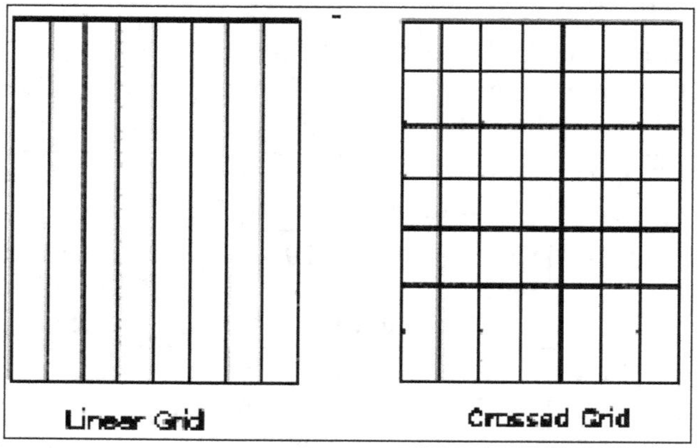

Figure 7.17: Linear and Cross Grids.

In general, the crossed grid will remove more scattered radiation than a linear grid of ratio equal to the combined ratios of its two parts, *e.g.*, a crossed grid, each of whose parts has 5:1 ratio, will remove more scattered radiation than a linear grid of 10:1 ratio. This advantage is more striking at voltages under 100 KVP.

The advantage of the linear grid over the crossed grid is that it may be used in tilted-tube techniques without undue "cut-off" in the radiograph. This is true with grid ratios 8:1 and lower and only if the angle of tilt of the tube is in a direction parallel to the length of the leads.

Tilting the tube at an angle across the leads will result in serious density reduction (cut off) on the film. With higher ratio grids, tube angling must be slight or focal distance long to avoid marked density variation.

7.3 Ultrasound Detection

Ultrasonic waves are generated by the application of an external electric field to a piezoelectric material. A previously derived inhomogeneous wave equation indicates that a gradient in eE (the product of the piezoelectric stress constant and the electric field) is the source for the generation of mechanical disturbances. Usually the surfaces of a piezoelectric material offer the sharpest discontinuity in both e and E, hence they are the strongest sources of sound. This can be shown by the following simplified figure. Assuming a parallel plate capacitor containing a piezoelectric material of dielectric constant ε.

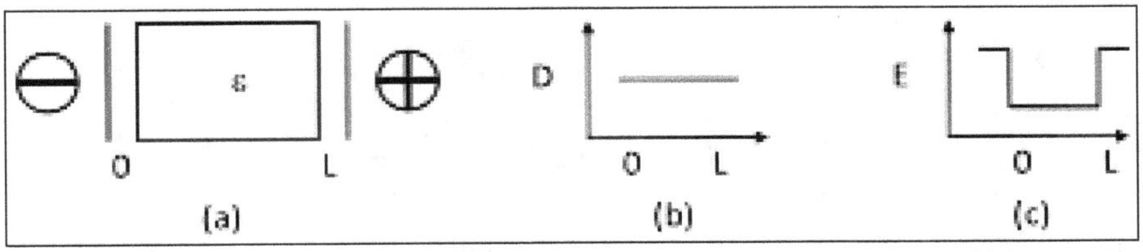

Consider the application of a charge density to the capacitor shown in figure (a). Since there are no free charge between the plates, we have $\frac{\partial D}{\partial z} = 0$ and hence D is a constant as illustrated in the figure (b). Additionally, the permittivity is typically higher inside the piezoelectric material, thus making the electric field E smaller than E in the outside. Therefore, figure (c) can be used to represent E and it becomes apparent that the only gradient in the electrical field occurs at the surfaces of the material. Consequently, the surfaces of the piezoelectric material are the predominant sources for the generation of ultrasound.

Piezoelectric detection of ultrasonic waves is reciprocal to the process of wave generation. In other words, the conversion of mechanical energy into electrical energy is also a phenomenon dominated by the behavior at the surfaces of the piezoelectric material. As shown below, the voltage measured across a piezoelectric plate is the integral of the electric field over the thickness of the crystal

$$V(t) = \int_0^L E(z,t)dz$$

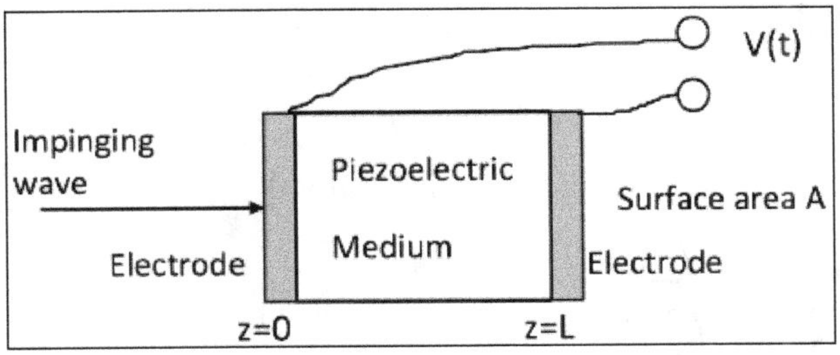

Using the constitutive relation relating the electric field to the strain and the electric displacement, the voltage becomes

$$V(t) = -\int_0^L \frac{e}{\varepsilon} S(z,t)dz + \int_0^L \frac{D(z,t)}{\varepsilon} dz$$

From Gauss's law, we have $D(z,t) = \frac{q(t)}{A}$ where $q(t)$ is the total charge on area A. Therefore, the above equation reduces to

$$V(t) = -\frac{e}{\varepsilon}\int_0^L \frac{\partial w(z,t)}{\partial z}dz + \frac{q(t)}{\varepsilon A/L}$$
$$\equiv -\frac{e}{\varepsilon}\left(w(L,t) - w(0,t)\right) + \frac{q(t)}{C_0}.$$

Where, $C_0 \equiv \frac{\varepsilon A}{L}$ is the capacitance of the piezoelectric plate.

The magnitude of the net displacement can be calculated from the boundary conditions placed at the surfaces of the material. In addition, the above equation shows that under open circuit situations, *i.e.*, $q(t)$ is not time-varying, the voltage developed across the plate is directly related to the relative displacements of the front and back surfaces of the material. If the thickness of the plate corresponds to an odd integral number of half-wavelengths of the ultrasonic wave impinging on the material (*i.e.*, the two surfaces oscillate 180° out of phase), then the relative displacement of the front and back surfaces, $w(L,t) - w(0,t)$, is the largest. In contrast, if the thickness of the crystal is an even number of half-wavelengths, then the amplitude of oscillation of the two surfaces is in phase, and therefore, $w(L,t) - w(0,t) = 0$.

Based on the piezoelectric detection equation that we previously derived, we have,

$$V(t) = -\frac{e}{\varepsilon}\left(w(L,t) - w(o,t)\right) + \frac{q(t)}{C_0}.$$

Taking the partial derivative with respect to time on both sides, we obtain

$$i(t) = C_0\frac{\partial V(t)}{\partial t} + C_0\frac{e}{\varepsilon}\left(u(L,t) - u(0,t)\right).$$

which can be represented by the following equivalent circuit.

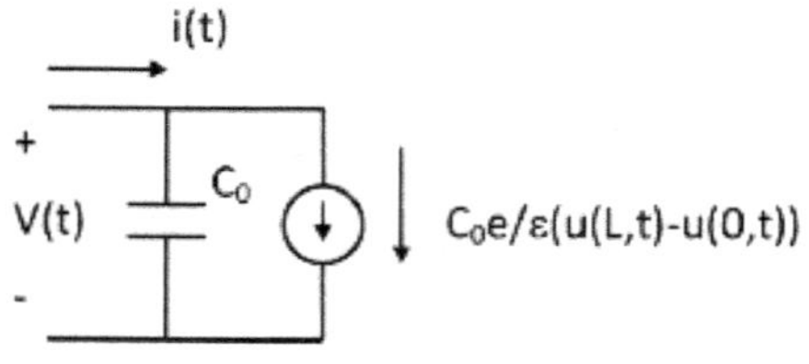

Defining force as we do voltage in electrical circuits, and particle velocity as we do current in electrical circuits, the non-piezoelectric component can be described using the other constitutive relation (*i.e.*, $T(z,t)_{piezo} = T(z,t)_{nonpiezo} + \frac{e \cdot q(t)}{\varepsilon \cdot A}$) and represented by the following circuit (a transmission line).

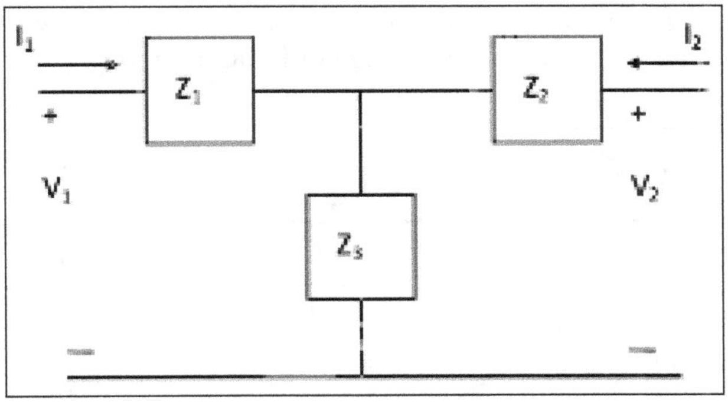

where $Z_1 = Z_2$ due to symmetry. Let Z_0represent the characteristic impedance of the piezoelectric material, we can obtain the following equations using methods similar to those used by deriving the acoustic wave equations

$$V(z,\omega) = V_1(\omega)e^{-j\omega z/c} + V_2(\omega)e^{j\omega z/c}$$

$$I(z,\omega) = \frac{1}{Z_0}\left(V_1(\omega)e^{-j\omega z/c} - V_2(\omega)e^{j\omega z/c}\right).$$

In order to obtain Z_1 and Z_3, the following relations need to be used.

$$Z_3 = \frac{V_2}{I_1}\bigg|_{I_2=0} = \frac{V(L,\omega)}{I(0,\omega)}\bigg|_{I(L,\omega)=0}$$

$$Z_1 + Z_3 = \frac{V_1}{I_1}\bigg|_{I_2=0}$$

Letting $I_2 = 0$, the following equation yields

$$Z_0 I(L,\omega) = 0 = V_1(\omega)e^{-j\omega L/c} - V_2(\omega)e^{j\omega L/c}.$$

Therefore,

$$V_2(\omega) = V_1(\omega)e^{-2j\omega L/c}.$$

and

$$Z_3 = Z_0\frac{V_1(\omega)\left(e^{-j\omega L/c} + e^{-j\omega L/c}\right)}{V_1(\omega)\left(1 - e^{-2j\omega L/c}\right)} = \frac{Z_0}{j\sin 2\pi L/\lambda} = -jZ_0 \operatorname{cosec} 2\pi L/\lambda$$

Furthermore, Z_1 (= Z_2) can be obtained as the following:

$$Z_1 = \frac{V(0,\omega)}{I(0,\omega)}\bigg|_{I_2=0} - Z_3 = Z_0\frac{\cos 2\pi L/\lambda - 1}{j\sin 2\pi L/\lambda} = jZ_0 \tan \pi L/\lambda$$

By using a transformer to couple the electrical components with the acoustic components, we then obtain the following equivalent circuit.

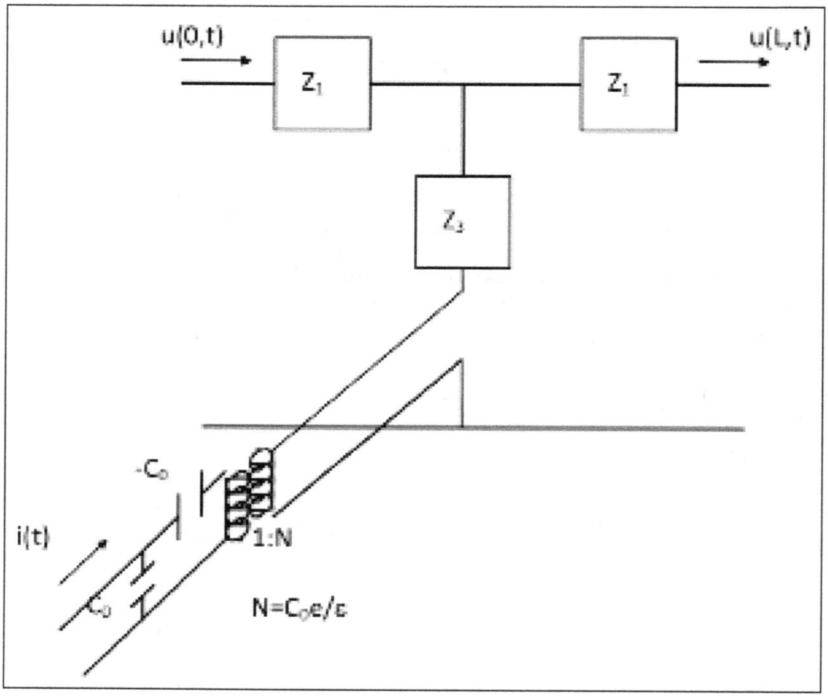

With the front and back faces of the transducer being loaded by mechanical impedances Z_F and Z_B, the equivalent circuit can be re-drawn as the following:

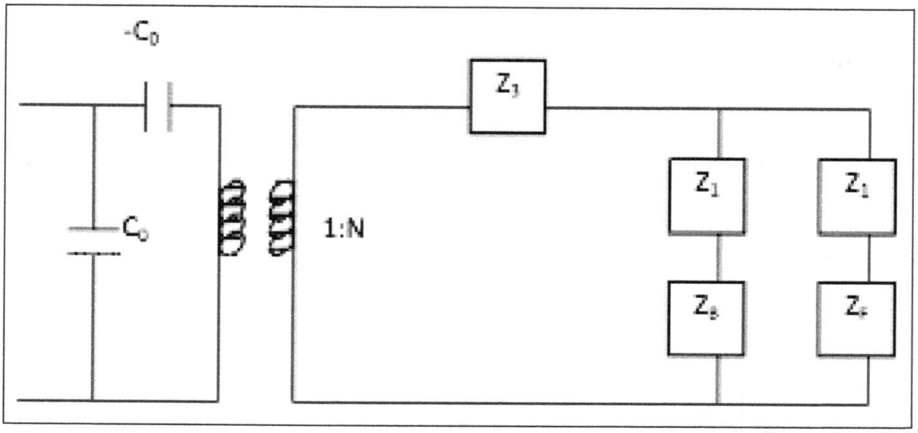

The above circuit can be further transformed to the following circuit identity with $Z_4=1/Z_1$.

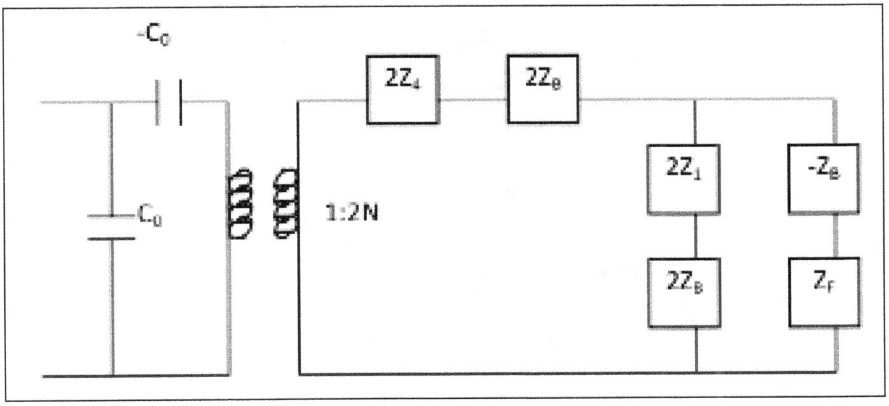

In the neighborhood of a mechanical resonance, impedance of the branch containing $2Z_1$ is large and the branch can be neglected. Therefore, the model can be approximated with the following circuit. Such an approximation is adequate to describe the operation of a transducer near resonance driving a wide variety of mechanical loads.

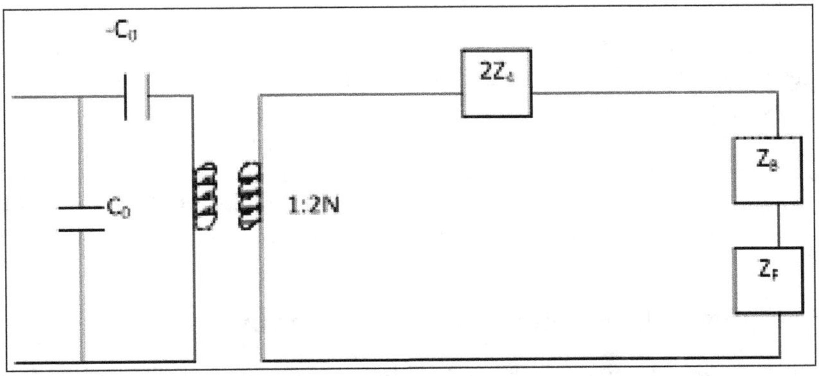

An alternative equivalent circuit which is the KLM *circuit* has been developed to be better suited for broadband operations.

7.4 Computed Tomography

Computed tomography (CT) is a diagnostic imaging test used to create detailed images of internal organs, bones, soft tissue and blood vessels. The cross-sectional images generated during a CT scan can be reformatted in multiple planes, and can even generate three-dimensional images which can be viewed on a computer monitor, printed on film or transferred to electronic media. CT scanning is often the best method for detecting many different cancers since the images allow your doctor to confirm the presence of a tumor and determine its size and location. CT is fast, painless, noninvasive and accurate. In emergency cases, it can reveal internal injuries and bleeding quickly enough to help save lives.

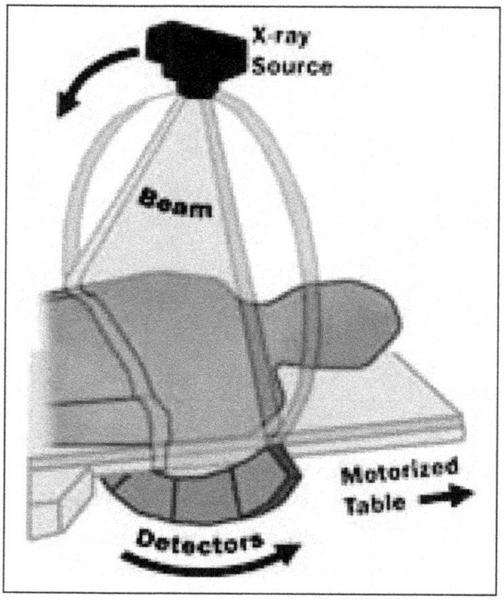

Figure 7.18: C.T Scan.

Computed tomography (CT) (Figure 7.18) is a way of using X-rays to take pictures or images in very fine slices through the part of the body that the doctor has asked to be investigated. When CT scanners were first invented, they took one slice at a time and were quite slow when compared to today's machines. Most modern scanners now take more than one slice at a time. This may range from 4 to 320 slices and up to 640 slices for the most recent machines. This is referred to as "multi-slice" or "multi-detector" technology, and may be abbreviated as MSCT or MDCT.

7.5 Principles of Magnetic Resonance Imaging

Magnetic resonance imaging (MRI) makes use of the magnetic properties of certain atomic nuclei. An example is the hydrogen nucleus (a single proton) present in water molecules, and therefore in all body tissues. The hydrogen nuclei behave like compass needles that are partially aligned by a strong magnetic field in the scanner. The nuclei can be rotated using radio waves, and they subsequently oscillate in the magnetic field while returning to equilibrium. Simultaneously, they emit a radio signal. This is detected using antennas (coils) and can be used for making detailed images of body tissues.

Unlike some other medical imaging techniques, MRI does not involve radioactivity or ionizing radiation. The frequencies used (typically 40-130 MHz) are in the normal radiofrequency range, and there are no adverse health effects. Very detailed images can be made of soft tissues such as muscle and brain. The MR signal is sensitive to a broad range of influences, such as nuclear mobility, molecular structure, flow and diffusion. MRI is consequently a very flexible technique that provides measures of both structure and function.

Magnetic Field Gradients

It is known that the basic equation of magnetic resonance is Larmor equation, $\omega = \gamma B_0$. In an NMR experiment a measurement of the frequency of precession of the magnetization gives information on the field experienced by that group of spins. By manipulating the spatial variation of the field in a known way, this frequency information now yields spatial information.

Consider a linear field gradient in B which increases along the x axis, such that

$$B = (B_0 + Gx)\mathrm{k}.$$

Where, G is the gradient strength.

This makes the Larmor equation $\omega(x) = \gamma(B_0 + Gx)$ or in its more general three dimensional form

$$\omega(r) = \gamma(B_0 + Gr)$$

Under a linear field gradient along the x axis, all the spins which lie at a particular value of x will precess at the same frequency. The FID from such a sample will contain components from each of the x values represented by the sample, and the frequency spectrum will therefore represent the number of spins that lie along that plane

$$\rho(\omega) = \rho(x) = \int \rho(x, y)\, dy$$

as shown in Figures 7.19.

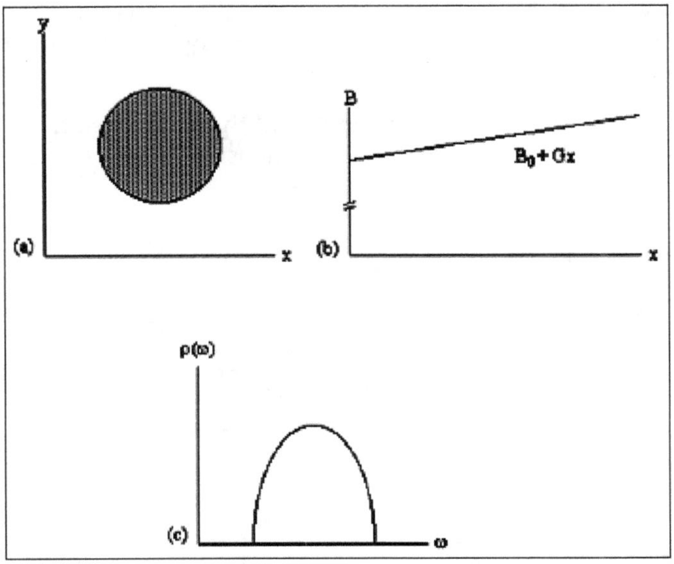

Figure 7.19: (a) A cylindrical object aligned along the z-axis. (b) Linear field gradient applied along the x-axis. (c) Plot of the number of spins at frequency w.

This simple spectrum therefore gives the spatial information about the object being imaged along one dimension. To build up the complete 3D image it is necessary to apply time varying field gradients. A number of methods for doing this are described later in this section, but first the notion of k-space is introduced, which is useful in describing all these techniques.

Reciprocal (k) Space

Having denoted the number of spins at a particular location r, as the spin density, r(r), the signal from the sample can be written

$$S(t) = \iiint \rho(\mathbf{r}) \exp[i\gamma\mathbf{G}\cdot\mathbf{r}t]d\mathbf{r}$$

The reciprocal space vector is defined as

$$\mathbf{k} = \gamma\mathbf{G}t$$

and the Fourier relationship between signal and spin density becomes obvious

$$S(\mathbf{k}) = \iiint \rho(\mathbf{r}) \exp[i\mathbf{k}\cdot\mathbf{r}]d\mathbf{r}$$

$$\rho(\mathbf{r}) = \iiint S(\mathbf{k}) \exp[-i\mathbf{k}\cdot\mathbf{r}]d\mathbf{k}$$

Thus k is the conjugate variable for r. The resolution of the image depends on the extent of k-space that is sampled, and it is by looking at how different imaging techniques cover k-space that they can usefully be compared (Figure 7.20). For example a gradient along the x-axis was applied to a cylindrical sample. This meant that values in k_x could be sampled, but not in k_y and so the complete 2D structure could not be obtained.

There are imaging techniques which sample all three dimensions of k space, but most techniques reduce the problem to two dimensions by applying slice selection.

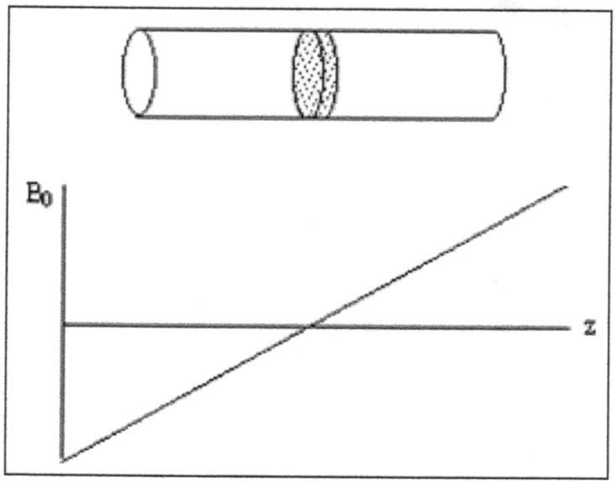

Figure 7.20: A Long Cylindrical Object Aligned Along the z-axis in a Field Gradient which Increases Linearly with Increasing z.

Slice Selection

Slice selection is a technique to isolate a single plane in the object being imaged, by only exciting the spins in that plane. To do this an r.f. pulse which only affects a limited part of the NMR spectrum is applied, in the presence of a linear field gradient along the direction along which the slice is to be selected as shown. This results in the excitation of only those spins whose Larmor frequency, which is dictated by their position, is the same as the frequency of the applied r.f. pulse.

Consider an r.f. pulse of duration 2t which is applied in the presence of a gradient along some axis, say z, such that $2tgB_1$= p/2 (a 90 degree pulse). The spins where $B=B_0$ (*i.e.* z=0) will precess into the transverse plane, whilst those where $B>>B_0$ (*i.e.* z>>0) will precess about B_{eff}=Dw/g k+B_1i, and have little effect on the transverse magnetisation as shown in Figure 7.21.

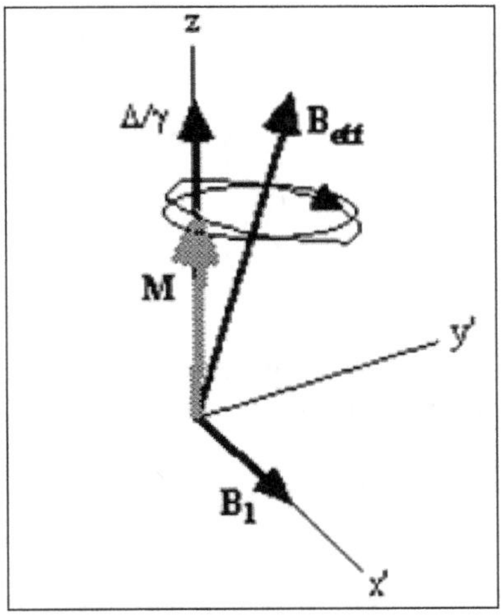

Figure 7.21: Effect of a 90 degree Pulse on Off-resonant Spins and Near-Resonant Spins.

To tailor the shape of slice selected, it is necessary to modulate the pulse profile. For small flip angles the relationship between the pulse modulation and slice profile can be derived. To do this, consider applying an r.f. pulse lasting from t=-T to +T, to a sample in a gradient G_z. The Bloch equations as shown in equation above in rotating frame, so by neglecting T_1 and T_2 relaxation, become

$$\frac{dM_x}{dt} = \gamma M_y G_z z$$

$$\frac{dM_y}{dt} = \gamma (M_z B_1(t) - M_x G_z z)$$

$$\frac{dM_z}{dt} = -\gamma M_y B_1(t)$$

Now these are transformed to another frame of reference which rotates at angular frequency gG_zz. Each slice will have a different reference frame, depending on the value of z, but considering just one of these gives

$$\frac{dM_{x'}}{dt} = -\gamma M_{z'} B_{y'} = -\gamma M_{z'} B_1(t)\sin\left[\gamma G_z z(t+T)\right]$$
$$\frac{dM_{y'}}{dt} = \gamma M_{z'} B_{x'} = -\gamma M_{z'} B_1(t)\cos\left[\gamma G_z z(t+T)\right]$$
$$\frac{dM_{z'}}{dt} = \gamma\left(M_{x'} B_{y'} - M_{y'} B_{x'}\right)$$

With the two reference frames coinciding at t=-T. If the assumption is made that the flip angles are small then it is possible to say that $M_{z'}$ varies very little, and takes the value M_0 so equations will become

$$\frac{dM_{x'}}{dt} = -\gamma M_{z'} B_{y'} = -\gamma M_{z'} B_1(t)\sin\left[\gamma G_z z(t+T)\right]$$
$$\frac{dM_{y'}}{dt} = \gamma M_{z'} B_{x'} = -\gamma M_{z'} B_1(t)\cos\left[\gamma G_z z(t+T)\right]$$
$$\frac{dM_{z'}}{dt} = \gamma\left(M_{x'} B_{y'} - M_{y'} B_{x'}\right)$$

If M_x and M_y are treated as real and imaginary parts of the complex magnetization M'_+, the equation of motion becomes

$$\frac{dM'_+}{dt} = i\gamma M_0 B_1(t)\exp\left[i\gamma G_z z(t+T)\right]$$

By integrating and converting back to the original (rotating) frame of reference, since

$$M_+(T) = M'_+(T)\exp\left(-i\gamma G_z z 2T\right)$$

gives

$$M_+ = i\gamma M_0 \exp\left[-i\gamma G_z zT\right]\int_{-T}^{T} B_1(t)\exp\left[i\gamma G_z zt\right]dt$$

This is the Fourier transform of the r.f. pulse profile, so a long pulse gives a narrow slice and vice-versa. This equation also shows one other thing about the magnetization in the slice after the pulse. There is a phase shift through the slice of gG_zzT which will cause problems for imaging later. This can be removed by reversing the gradient G_z for a duration T, which is known as slice refocusing.

To excite a sharp edged slice, the r.f. pulse modulation required is a sin c function. For practical implementation of this kind of pulse it is necessary to limit the length of the pulse, and so usually a five lobe sin c function is used. The above approximation is only correct for small pulse angles. For larger flip angles, it is hard to analytically determine the pulse modulation required for a desired slice profile, and so it is usually necessary to use some form of iterative method to optimize pulses.

Using such techniques, it is possible to design a range of pulses which are able to select different slice profiles and carry out other spin manipulation techniques such as fat or water suppression. It is also possible to design pulses which do not require refocusing. Usually the limit in such novel designs is their practical implementation, and so the most common three pulse modulations used are a 'hard' pulse, with top-hat modulation, a Gaussian modulation or a 'soft' pulse, with truncated sin c modulation.

If any slice other than the central ($B=B_0$) slice is required, then the frequency of oscillation can be altered and the slice selected will be at the position along the z-axis given by

$$Gz = \frac{\omega}{\gamma} - B_0$$

Early MR Imaging Techniques

The extensively used techniques in MRI are fourier based here the spin-warp technique and Echo Planar Imaging (EPI) are used. The f.o.n.a.r. (*f*ield f*o*cused *n*uclear m*a*gnetic *r*esonance) technique was proposed by Damadian, and was used to produce first whole body image. The basis of technique is to create a shaped B_0 field which has a central homogeneous region, surrounded by a largely inhomogeneous region. The signals from the inhomogeneous regions will have a very short T_2^* and can thus be distinguished from the signal from the homogeneous region. By scanning the homogeneous field region across the whole of the sample an image can be made up. This is a time consuming procedure taking tens of minutes for a single slice. The two advantages of the technique are its conceptual simplicity and the lack of a requirement for the static field to be homogeneous over a large area. Shaped r.f. pulses can also be used to isolate a small region, and such methods form the basis of localized spectroscopy.

If then the three dimensional samples can be reduced to a set of one dimensional components then the whole sample can be imaged. This can be done by selective irradiation, or slice selection, in two of the dimensions. First, in the presence of a gradient along the z-axis, a selective pulse is applied which saturates all the spins outside the plane of interest. Then a gradient is applied along the x-axis, and those spins not saturated are tipped into the transverse plane by a selective 90 degree pulse.

Immediately after the second r.f pulse the only region with any coherent transverse magnetization is the line of intersection of the two selected planes. A gradient is now applied along the y-axis, and the evolution of the FID recorded (Figure

7.22). Fourier transform of the FID gives the proton densities along that line. By repeating the line selection for all the lines in the plane, an image of the whole of the plane can be built up. There are many variations of the line scan technique, some of which utilize 180 degree pulses, but they are inefficient in comparison to the Fourier methods.

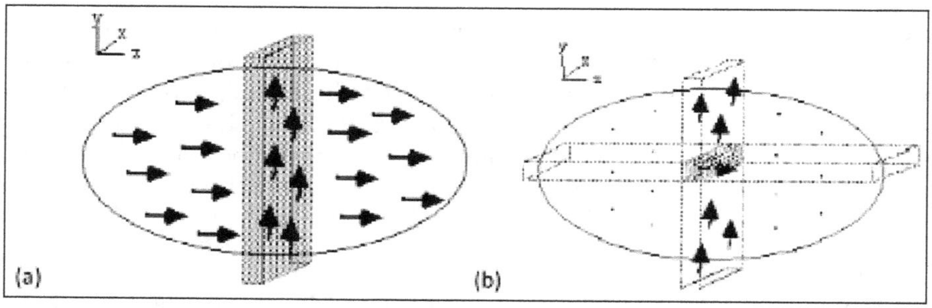

Figure 7.22: Line Selection using Two rf Pulses.

One final method of interest is projection reconstruction. This is the method used to build up X-ray CT scans, and was the method used to acquire the early MR images. Following slice selection, a gradient is applied along the x-axis and the projections of the spin densities onto that axis obtained by Fourier transformation of the FID. Then a linear gradient is applied along an axis at some angle to the x-axis, q. This can be achieved by using a combination of the x and y gradients $G_y = G\sin\theta$ $G_x = G\cos\theta$.

Projections are taken as q is incremented up to 180 degrees. The set of projections can then be put together using back projection. This distributes the measured spin densities evenly along the line normal to the axis it was acquired on. By reconstructing all the angled projections the image appears. There is however a blurred artefact across the whole image, where an attempt was made to assign spin density to areas where there is in fact none, and star artefacts where a finite number of projections have been used to define point structures. This can be corrected using a technique called filtered back projection, which convolves each of the profiles with a filter (Figure 7.23).

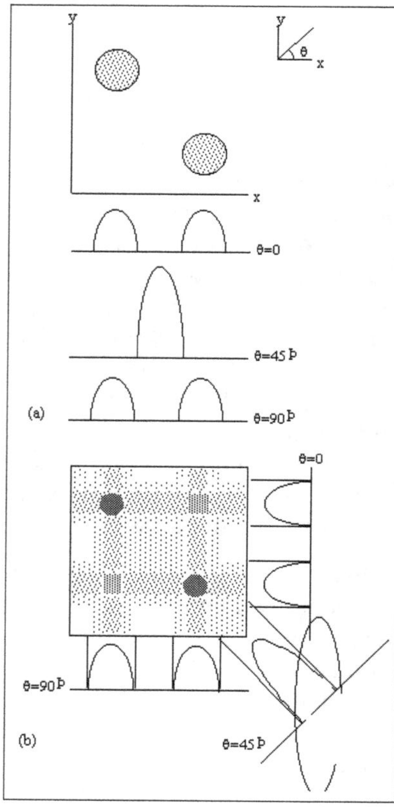

Figure 7.23: The Technique of Projection Reconstruction.

$$\frac{1}{t(r - r')^2}$$

Fourier and Echo Planar Imaging

The Fourier zeugmatography sequence can be split into three distinct sections, namely slice selection, phase encoding, and readout. The pulse sequence diagram for the sequence is shown in Figure 7.24. Such diagrams are commonly used to describe the implementation of a particular MR sequence, and show the waveform of the signal sent to the three orthogonal gradient coils and the r.f. coil.

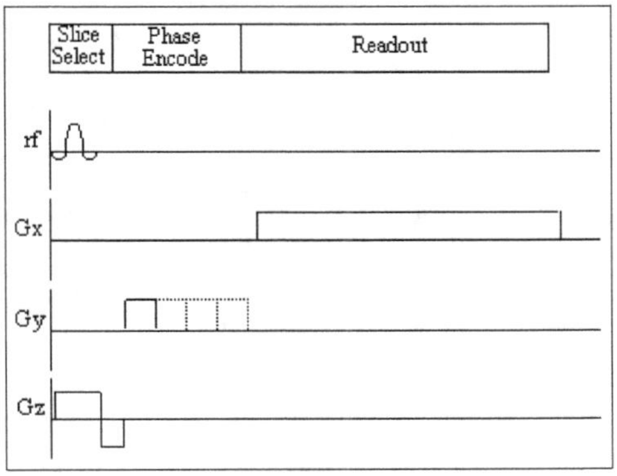

Figure 7.24: Pulse Sequence Diagram for the Fourier Zeugmatography Imaging Technique.

Having excited only those spins which lie in one plane, a gradient is applied along the y-axis. This will cause the spins to precess at a frequency determined by their y position, and is called phase encoding. Next a gradient is applied along the x-axis and the FID is collected. The frequency components of the FID give information of the x-position and the phase values give information of the y-position (Figure 7.25)

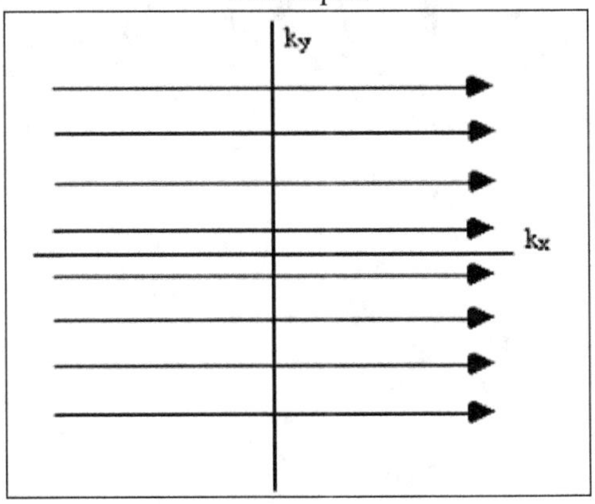

Figure 7.25:. K-space Sampling with Fourier Zeugmatography.

More specifically, if a gradient of strength G_y is applied for a time t_y during the phase encoding stage, and then a gradient G_x is applied for a duration t_x the signal recorded in the FID is given by

$$S(t_x, t_y) = \iint \rho(x,y) \exp\left[i\,\gamma(G_x x t_x + G_y y t_y)\right] dx\, dy$$

By writing $k_x = \gamma G_x t_x$ and $k_y = \gamma G_y t_y$ equation (2.51) becomes

$$S(k_x, k_y) = \iint \rho(x,y) \exp\left[i(k_x x + k_y y)\right] dx\, dy$$

so the single step is equivalent to sampling one line in the k_x direction of k-space. To cover the whole of k-space it is necessary to repeat the sequence with slightly longer periods of phase encoding each time.

Having acquired data for all values of k_x and k_y, a 2D Fourier transform recovers the spin density function (Figure 7.26).

$$\rho(x,y) = \iint S(k_x, k_y) \exp\left[-i(k_x x + k_y y)\right] dx\, dy$$

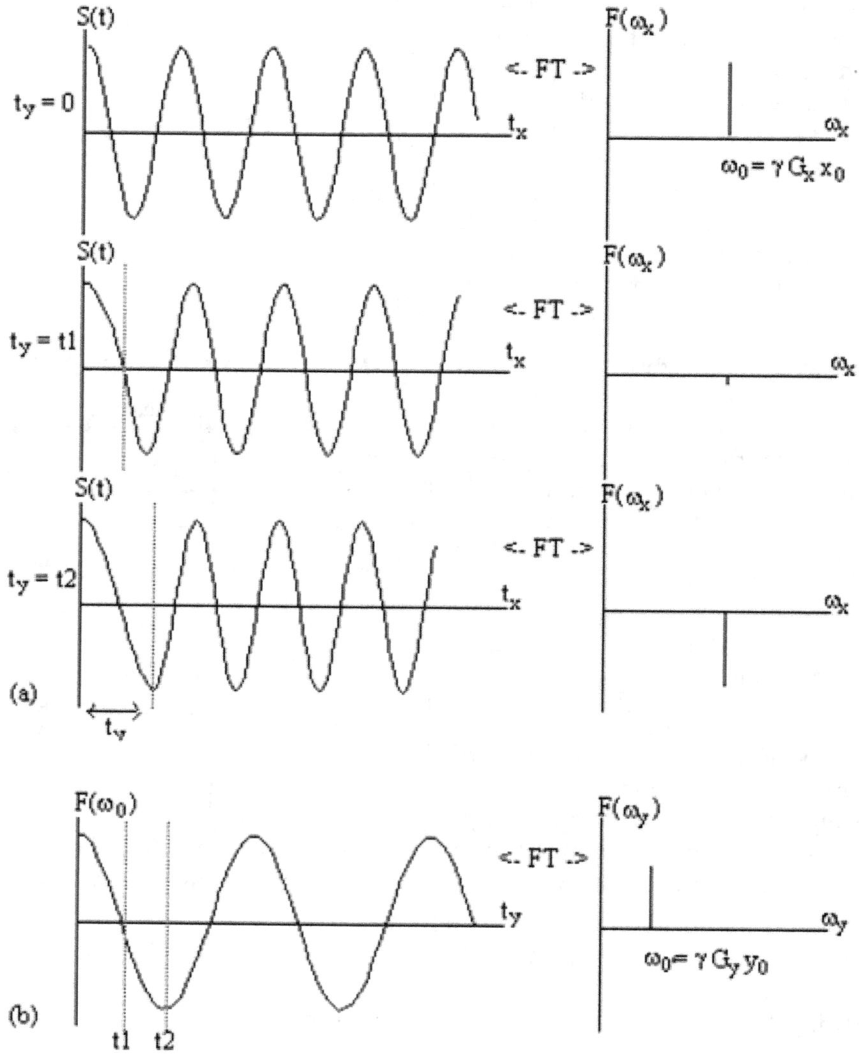

Figure 7.26: Two Dimensional Fourier Image Reconstruction following Fourier Zeugmatography.

One drawback of this technique is that the time between exciting the spins, and recording the FID varies throughout the experiment. This means that the different lines in the k_y direction will have different weighting from T_2^* magnetization decay.

This is overcome in spin-warp imaging by keeping the length of the y gradient constant for each acquisition, and varying k_y by changing the gradient strength. The pulse diagram for this technique is shown in Figure 7.27.

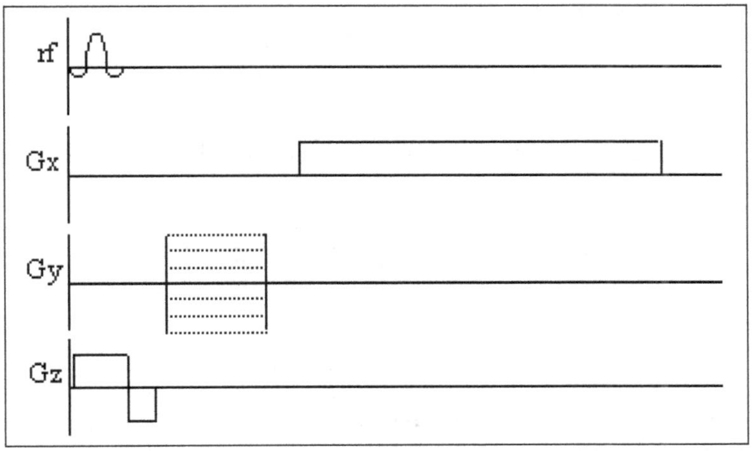

Figure 7.27: Pulse Sequence Diagram for the Spin-Warp Technique. Phase encoding is performed by varying the magnitude of the y gradient.

It is desirable to have as much signal as possible for each FID, and a necessity that the amount of transverse magnetization available immediately after the r.f. pulse is the same for each line. This can be a problem since the recovery of longitudinal magnetization is dependent on spin-lattice relaxation, and T_1 values in biomedical imaging are of the order of seconds. Keeping the time between adjacent spin excitations, often known as TR, the same throughout the image acquisition will keep the transverse magnetization the same for each FID, provided the first few samples are discarded to allow the system to come to a steady state.

Leaving the magnetization to recover fully, however, would be very costly in time, so it is usually necessary to have a TR which is less than T_1. To maximize the signal received for small TR values it is possible to use a smaller flip angle than 90 degrees. The transverse magnetization that is available after such a pulse is less than it would be after a 90 degree pulse, but there is more longitudinal magnetization available prior to the pulse. To optimize the flip angle q, for a particular TR, we first assume that the steady state has been reached, that is that

$$M_Z(0) = M_Z\,(T_R) = \mathrm{M}'$$

Now the magnetization is flipped by a q degree pulse, and the z-magnetization becomes

$$M_Z(0') = M'\cos\theta.$$

The recovery of the magnetization is governed by the equation

$$\frac{dM_Z}{dt} = -\frac{M_o - M_Z}{T_1}$$

which can be integrated to find M′,

$$M' = M_0 \frac{1 - \exp(-T_R/T_1)}{T - \cos\theta \exp(-T_R/T_1)}$$

The transverse magnetization following the pulse, which we want to maximize, is given by

$$M_y = M' \sin\theta = M_0 \frac{1 - \exp(-T_R/T_1)}{T - \cos\theta \exp(-T_R/T_1)} \sin\theta$$

which has its maximum value when

$$\cos\theta = \exp(-T_R/T_1)$$

The angle this occurs at is known as the Ernst angle. The amount of signal available is very dependent on the repetition time TR. For example, if a sample has a T_1 of 1s, then at a TR of 4s $M'=0.98M_0$, however as TR is reduced to 500ms, $M'=0.62M_0$.

If a 3D volume is to be imaged then it is possible to acquire extra slices with no time penalty. This is because it is possible to excite a separate slice, and acquire one line of k-space, whilst waiting for the longitudinal magnetization of the previous slice to recover. This technique is called multi-slicing.

The common implementation of the spin-warp technique, FLASH (Fast Low-Angle SHot imaging), uses very small flip angles (~5 degrees) to run at with a fast repetition rate, acquiring an entire image in the order of seconds.

In Echo Planar Imaging (EPI) the whole of k-space is acquired from one FID. This is possible because, having acquired one set of frequency information, the sign of the readout gradients can be reversed and the spins will process in the opposite direction in the rotating frame and subsequently rephase causing a regrowth of the NMR signal. This is a called a gradient echo (Figure 7.28). By switching the readout gradient rapidly, the whole of k-space can be

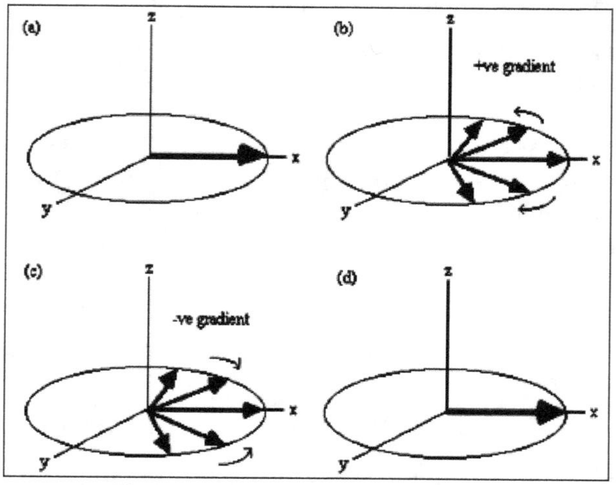

Figure 7.28: Magnetisation Vector in Rotating Frame under Gradient Echo.

sampled before spin-spin (T_2) relaxation attenuates the transverse magnetization. Phase encoding is again used in order to sample k_y.

The pulse sequence diagrams and k-space trajectories for EPI are shown in Figure 7.29.

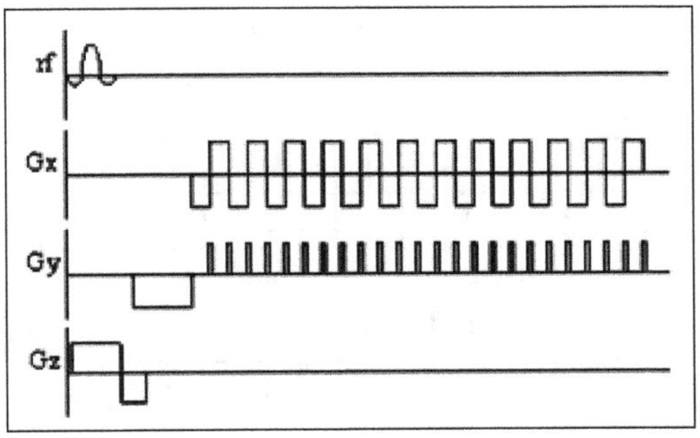

Figure 7.29: (a) Pulse sequence diagram and (b) k-space sampling pattern for echo planar imaging.

The three gradients in EPI are usually labelled the slice select (z), blipped (y) and switched (x), because of their respective waveforms. Echo planar imaging is a technically demanding form of MRI, usually requiring specialized hardware, however it has the advantage of being a very rapid imaging technique, capable of capturing moving organs like the heart, and dynamically imaging brain activation. This is only a brief introduction to EPI, since its strengths and limitations are discussed in more detail in the following chapters.

Other Imaging Sequences

There are numerous variations on the basic MRI sequences described above. Several of them, notably interleaved EPI. Other important sequences are outlined here.

It was shown how phase encoding enabled the information on the second dimension to be added to the one dimensional line profile. It is possible to extrapolate this procedure to the third dimension by introducing phase encoding along the z axis. Thus the 2D-FT technique is extended to a 3D-FT technique. All such volumar imaging sequences first involve the selection of a thick slice, or slab. Then phase encoding is applied in the z-direction and the y-direction, followed by a readout gradient in the x-direction, during which the FID is sampled. The assembled FID's are then subject to a three dimensional Fourier transform yielding the volume image. Phase encoding can also be used in EPI instead of multi-slicing. The slice select gradient and r.f. pulse being replaced by a slab selective pulse and a phase encoding gradient along the z-axis (Figure 7.30).

Going even further, it is possible to acquire all the data to reconstruct a 3D volume from one FID, in the technique called Echo Volumar Imaging (EVI). This uses another blipped gradient in the z direction, as shown in Figure 7.30. The limitation in EVI is the need to switch the gradients fast enough to acquire all the data, before T_2^* destroys the signal.

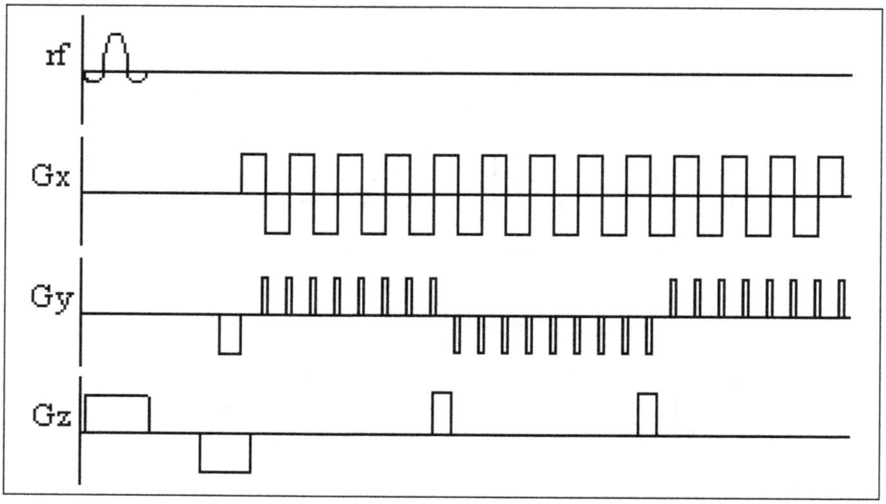

Figure 7.30: Pulse Sequence Diagram for Echo Volumar Imaging.

Often the main limitation in implementing fast imaging sequences such as EPI is switching the gradients at the fast rates required. A sequence which is similar to EPI, but slightly easier to implement is spiral imaging. This covers k-space in a spiral from the centre outwards, which requires sinusoidal gradients in x and y, increasing in amplitude with time. Such gradient waveforms are easier to produce than the gradients required for EPI. Spiral imaging also has the advantage of sampling the centre of k-space first, and so the low spatial frequencies, that affect the image the most are sampled first, when the signal has not been eroded by T_2^*. The pulse sequence diagram and coverage of k-space for spiral imaging are shown in Figure 7.31.

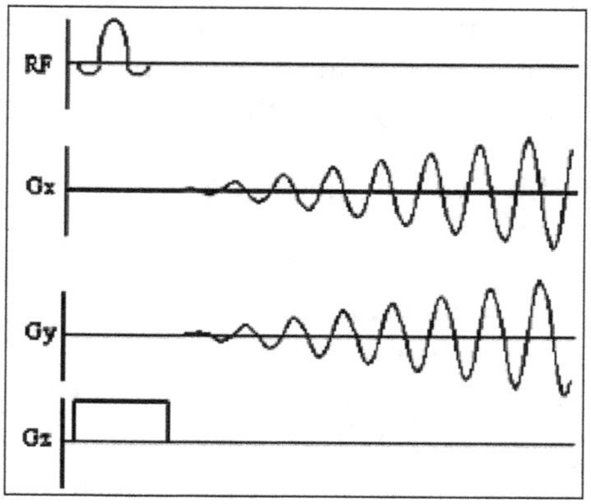

Figure 7.31: Pulse Sequence Diagram and k-space Sampling Diagram for Spiral Imaging.

7.6 Introduction of Molecular Imaging

The field of molecular imaging emerged in the early 1990's as scientists from multiple disciplines, including cell biology, biomedical engineering, chemistry, mathematics, medicine, pharmacology and genetics began working toward the development of imaging instruments, imaging probes, assays, and quantification techniques to elucidate molecular mechanisms in biology and medicine.

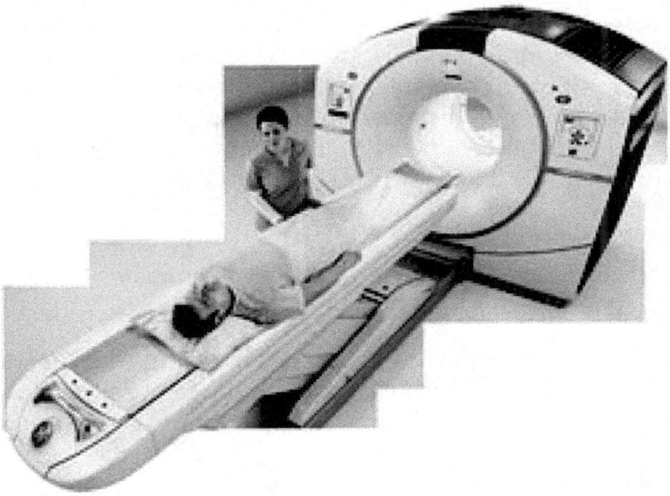

Molecular imaging aims to non-invasively visualize, characterize and quantify normal and pathologic processes within the living organism at the cellular and subcellular level. In contrast to traditional biomedical imaging by microscopy, in which excised tissues are typically examined to characterize histological changes and thus identify an underlying disease process, molecular imaging targets distinct molecular pathways in vivo, providing visual and quantitative information for diverse research applications.

For example, investigators have applied molecular imaging to:

1. Non-invasively characterize the stages and progression of a disease process and establish signature biomarkers;
2. Assess the efficacy of standard or experimental treatment modalities in small-animal models of human disease;
3. Characterize the trafficking of stem cells and immune cells;
4. Analyze the bio-distribution of drugs and the dynamics of drug/receptor interactions;
5. Investigate the cellular and subcellular basis of brain disorders;
6. Assess metabolic changes, particularly in the brain, heart and tumors;
7. Detect tissue hypoxia.

Because biological systems are, at one level, highly complex and dynamically changing assemblies of molecular structures and interactive pathways, molecular imaging provides the tools to investigate in real time the distribution and activity of targeted biomolecules as they undergo changes in space and time.

Molecular imaging offers scientists in many different specialties significant advantages over traditional research paradigms. First, molecular imaging procedures can be conducted in the living organism. Whereas studies of tumor responsiveness to a new therapeutic agent, for example, would have traditionally involved a large cohort of animals from which subsets would be analyzed histologically at multiple time points, molecular imaging allows characterization of tumor development and response to a therapeutic, and even response to discontinuation of the therapeutic, within the same small set of animals imaged longitudinally at multiple time points.

An associated advantages of molecular imaging:

1. Study be conducted using significantly less animals
2. Statistical power is increased as every animal serves its own control

Other advantages include the ability of molecular imaging procedures to interrogate the whole body, in addition to focusing on specific regions, and to visualize the molecular target of interest in 3-dimensional space. Finally, molecular imaging is becoming a key bridging technology to translate experimental preclinical findings into the clinical environment.

Imaging Instruments

These imaging modalities include magnetic resonance imaging (MRI), X-ray computed tomography (CT), positron emission tomography (PET), single photon emission computed tomography (SPECT), and optical (bioluminescence and fluorescence imaging). In addition, a limited band of the sound spectrum is utilized for molecular imaging by ultrasound. Beginning in the early 1990's, new imaging instruments were designed and fabricated with the aim of advancing the use of small research animals, particularly mice, in molecular imaging studies through the use of dedicated imaging instruments with higher sensitivity and resolution than their clinical counterparts. The continuing evolution of imaging technology has seen a rapid growth in the development and use of multimodality instruments, both in the preclinical and clinical settings.

Dual modality, or hybrid, PET/CT scanners have become the preferred system for clinical nuclear medicine, and dedicated PET/CT and SPECT/CT scanners for small animal imaging are also now used routinely. PET/MRI instruments also are being developed. The powerful advantage of these blended instruments derives from melding the functional imaging strength of one technology, such as PET and SPECT, with the high resolution anatomical imaging afforded by MRI or CT. Finally, new instruments including combined Optical/X-Ray and other technologies have been recent additions for small animal imaging. It can be anticipated that the growing use of molecular imaging in biology and medicine will drive the development of new imaging instruments as well as the refinement of our current imaging technologies.

Imaging Contrast Agents

Once a target molecule has been designated, an appropriate imaging contrast agent must be selected. The contrast agent is typically a molecule that can bind to the target molecule or be trapped enzymatically in a cellular compartment. In addition the imaging agent is labeled with a moiety that renders it visible to a particular imaging modality. Labels in common use include radio-nuclides (*e.g.*, ^{11}C, ^{18}F and ^{64}Cu for PET; ^{99m}Tc, ^{123}I and ^{111}In for SPECT), fluorescent molecules and paramagnetic ions. A number of small particles that can be covalently bound to targeting molecules have been developed, including nanoparticles, liposomes and micro-bubbles. These particles may be intrinsically fluorescent (*e.g.*, quantum dots), give strong ultrasound echoes or can be tagged with radioactive ions, fluorescent molecules or paramagnetic ions. Targeted molecular micro CT imaging is limited by the need for impractically high concentrations of contrast agent, but iodine-based contrast agents may be used to enhance soft tissue contrast and improve anatomical imaging. The development of new contrast agents for each imaging modality, and for multimodality imaging, is progressing rapidly as investigators seek new probes with the potential to characterize the stage of a disease and evaluate the response to therapy when used at diagnostic doses, as well as to deliver a molecular therapy when administered at higher doses.

7.7 Radiotracers and Imaging Radio-Nuclides

A radioactive tracer is a chemical compound in which one or more atoms have been replaced by a radioisotope. Monitoring its radioactive decay, a radiotracer can be used to explore the mechanism of chemical reactions. They are also used for flow visualization through different technologies, such as Single Photon Emission Computed Tomography, Positron Emission Tomography and Computed Radioactive Particle Tracking.

Radiotracer technology is playing a more and more important role in industry. It is used to diagnose specific causes of inefficiency in a plant or process operation and to generally investigate processes in industries and those related environments where a great cost-benefit ratio can be gleaned from process optimization and troubleshooting, such as in the transport of sediments. It is expected that this important role will continue to expand, especially if students and engineers are exposed in their academic training to the many possibilities of this tool for research, development and application.

Various requirements need to be met before running a radiotracer investigation. The most important of these is to select an appropriate radiotracer. It is of fundamental importance that the radiotracer compound behaves in the same way as the material to be traced. To obtain reliable and meaningful results, an industrial radiotracer must also meet other basic requirements, such as a suitable half-life and radiation energy, as well as physical and chemical stability. It must also be easily and unambiguously detectable.

Before injecting a tracer into a system it must be clear how it will behave inside it. In certain circumstances, the tracer may undergo decomposition, phase change, undesirable absorption and adsorption, or chemical interaction with system constituents with incorrect results.

Radiotracers are often used in conjunction with pharmacologic stressors, particularly for the detection of coronary artery disease. Recent evidence suggests that radiotracer pharmacokinetics may be influenced by the type of pharmacologic stress applied, hence biological models have an important role in evaluation of pharmacologic stressors.

The effects of pharmacologic stress on radiotracer kinetics can be initially evaluated in short-term anesthetized models. However, the effects of inotropic agents such as dobutamine and arbutamine on global hemodynamics, regional myocardial flow, and function may be greatly influenced by the type of anesthesia. For example, the anesthetized dog is more sensitive to inotropic stimulation; in contrast, it requires much higher doses of adenosine to elicit the same effect on coronary flow. These differences must be taken into consideration when designing experimental protocol involving a pharmacologic stressor. Ultimately, pharmacologic stressors must be evaluated in long-term, conscious experimental models to define their true effects on radiotracer kinetics.

Radiotracer Techniques

Radiotracer techniques use radio-nuclides as probes to quantify physiological processes, *e.g.* cerebral blood flow, or to label biochemical pathways or specific molecules. The probe is spatially localized by detecting the emitted radioactivity, while variations in local radioactivity over time can be used to identify rate constants and physiological fluxes. Radiotracers are sensitive and can be adapted to different uses, including marking substrates to investigate biochemical processes, labeling a drug target, or labeling the drug itself. Molecular tracers are the only way to measure receptors and their function because of their low concentrations (nano- and picomolar range).

Single-Photon Emission Computed Tomography

Single-photon emission computed tomography (SPECT) detects tracer molecules labeled with gamma-emitting radioisotopes. Typically it uses an array of two or three gamma cameras that rotate around the subject and is increasingly combined with a computed tomography (CT) system that offers improved spatial resolution and registration with structural images. SPECT is more widely available than PET, although the range of tracers for SPECT is more limited.

SPECT is commonly used for measuring cerebral perfusion (regional cerebral blood flow; rCBF) using, for example, ^{99m}Tc- or ^{123}I-labeled radio-pharmaceuticals.

Receptor imaging with SPECT is less common but is used in research with agents available for dopamine and serotonin transporter imaging (^{123}I-2β-carbomethoxy-3β-(4-iodophenyl)tropane; ^{123}I-β-CIT) and dopaminergic D_2 receptors (^{123}I-IBZM).

rCBF can be measured using freely diffusible tracers such as ^{133}Xe. The cerebral transit of the tracer is measured and, when combined with an estimate of the input function, a mathematical model can be used to estimate cerebral blood flow.[7] More common is the use of a so-called static tracer such as ^{99m}Tc-exametazime. Such tracers pass through the blood–brain barrier and are then retained by the brain for several hours. The accumulation of the tracer over time is proportional to the rCBF, although absolute quantitation is difficult as accumulation also depends on the mechanism of retention in the brain.

Positron Emission Tomography

PET uses radiotracers labeled with positron-emitting radioisotopes. It has a higher signal-to-noise ratio and better spatial resolution than SPECT. PET has an exquisite sensitivity allowing detection of tracers in the nano- to picomolar range. This allows the biochemical system to be probed without significant pharmacological effect, as well as with significant pharmacological effect if advantageous. PET has been used to map receptor systems, measure receptor occupancy by pharmacological agents, assay enzyme activity, measure cerebral oxygen and glucose metabolism and cerebral blood flow, and therefore to map task-specific functions in the human brain, including cognitive, drug-stimulated, motor, and sensory activity. As for SPECT, PET is normally combined in the same machine with CT to allow the collection of combined structural and functional information.

In general, a cyclotron is needed to generate a wide range of positron-emitting radio-nuclides. These radio-nuclides are used to label compounds of interest, which are normally introduced into the subject intravenously. Decay of the radionuclide results in the emission of a positron and a neutrino. The neutrino is not detected, while the positron interacts with electrons in the tissue finally being annihilated along with one of the electrons and liberating two gamma-rays, each with an energy of 511 keV. The gamma-rays are emitted approximately in opposite directions and are detected by rings of detectors arranged around the subject. The solid-state detectors look for coincident events in which gamma-rays are detected almost simultaneously on opposite sides of the head.

The line along which the positron annihilation took place can therefore be reconstructed. By combining data from all of the different angles, images can be reconstructed that reflect the concentration of the positron-emitting radionuclide. The higher signal-to-noise ratio of PET with respect to SPECT arises from the lack of a requirement for collimation at the detectors, reduced attenuation of the higher-energy gamma-rays, and the use of complete rings of detectors rather than a moving camera. It is possible to construct kinetic models of the PET tracers that, when combined with a time-series of PET images, can reveal the rates of biological reactions in which the tracers are engaged.

The wide range of biological measurements that can be made with PET is only possible because some of the most common biological elements can form positron-emitting isotopes, including carbon (^{11}C), nitrogen (^{13}N), and oxygen (^{15}O). Furthermore, in many molecules, fluorine-18 (^{18}F) can be used as an analog of hydrogen. Once the appropriate isotope is made, it must be incorporated into an organic molecule that has the required biological activity for investigating the biological process of interest.

Cerebral Metabolism and Blood Flow

The cerebral rate of metabolic oxygen consumption can be measured using ^{15}O-labeled O_2. However, the most commonly used PET radiotracer is ^{18}F-fluorodeoxyglucose (FDG), which is transported across the blood–brain barrier. In cells it is converted to FDG-6-phosphate, which is then trapped inside the cell. Knowing the input function of the tracer from arterial blood and with a time-series of FDG measurements, the cerebral metabolic rate for glucose (CMRGl) can be mapped. This has found application in the assessment and identification of brain tumors, which often show increases in aerobic and anaerobic glycolysis. FDG-PET can be used to characterize different types of neuro-degeneration that may result in dementia, including Alzheimer disease (AD).

In AD, reduced CMRGl is seen in the parietal and temporal lobes. These patterns of low metabolism can be used to help distinguish between AD and other dementias such as frontotemporal dementia.

Early studies of functional mapping of the human brain were performed using FDG-PET with subjects performing tasks and demonstrated increased CMRGl. This laid the foundations for the vastly expanded field of human functional brain mapping. In the domain of PET, FDG studies of brain activation have been largely superseded by cerebral blood flow measurements because of the opportunities to perform repeated measurements in the same individual.

Cerebral blood flow is measured using a diffusible tracer, *e.g.* ^{15}O-H_2O, which is therefore able to pass through the blood–brain barrier. The quantification of blood flow relies on entry of the tracer into the brain, which depends on the flow rate to the tissue and is not limited by the diffusion of the tracer into the tissue. A compartmental model can be applied describing the kinetics of ^{15}O-H_2O in which blood flow is one of the parameters. For studies aiming to map brain functions, relative cerebral blood flow (CBF) measurements are often made, relying, for short scans, on the near-linear relationship between blood flow and the distribution of ^{15}O-H_2O.

Glossary

- **Grid cassette:** It is a radiographic cassette with a grid permanently installed in it.
- **Grid cutoff:** It is an excessive loss of radiation beam as of incorrect angulation between the tube and lead strips.

- **Focused grid:** It is the lead strips in grid that are sloped slightly inwards at top so that the apertures more closely approximate the angle taken by the rays in a diverging beam.
- **Linear grid:** A cassette grid where radiopaque strips are parallel to each other.
- **Parallel grid:** It is the lead strips in grid that are parallel to each other and vertical to plane of film.
- **Grid ratio:** It is the ratio between the height of strips to distance separating them. The greater the ratio the more rays will be filtered out.
- **Reciprocating grid:** It is a grid that moves during the exposure of subject to beam of radiation.
- **Rotating grid:** It is a grid that moves while X-rays are being generated.
- **Stationary grid:** It is a grid that is stationary while the film is being exposed so that there is a pattern of grid lines on the radiograph.
- **Grid tube:** It is a special tube designed to take cineradiographic X-rays.

References

Cherry SR. In vivo molecular and genomic imaging: new challenges for imaging physics. Phys Med Biol; 49:R13-R48, 2004.

Cherry SR. Multimodality in vivo imaging systems: Twice the power or double the trouble? Annu Rev Biomed Eng; 8:35-62, 2006.

Gambhir SS. Just what is molecular imaging? mi gateway; 1:5-7, 2007.

Jaffer FA, Libby P, Weissleder R. Optical and multimodality molecular imaging; Insights into atherosclerosis. Arterioscler Thromb Vasc Biol; 29:1017-24, 2009.

Koo V, Hamilton PW, Williamson K. Non-invasive in vivo imaging in small animal research. Cellular Oncol; 28(4):127-139, 2006.

Lyons SK. Advances in imaging mouse tumour models in vivo. J Pathol; 205:194-205, 2005.

Rowland DJ, Cherry SR. Small-animal preclinical nuclear medicine instrumentation and methodology. Semin Nucl Med; 38:209-22, 2008.

Tsui BMW, Kraitchman DL. Recent advances in small-animal cardiovascular imaging. J Nucl Med; 50:667-70, 2009.

Weissleder R, Pittet MJ. Imaging in the era of molecular oncology. Nature; 452:580-89, 2008.

8

Radiation Protection

8.1 Neutron Thermalization in Hydrogenous Moderators

Neutrons produced by the fusor are all considered "Fast neutrons". All neutron sources tend to produce fast neutrons. All of the best detectors for neutrons rely on "thermal neutrons" to be detected by some secondary process. For detection of fusor produced and any man made source of neutrons, (fast), it is incumbent on the would-be neutron metrologist to slow these neutrons down to thermal velocities allowing detectors sensitive to these "slow neutrons" to count them.

Moderators can be made of a number of materials, but moderators made of hydrogen rich materials work best and take up the least amount of space. Moderation or "slowing" of neutrons occurs when neutrons effectively "hit" a proton within the hydrogen of the moderator causing the proton to recoil, much like a billiard ball. Of course, just like the billiard ball, the proton is sent recoiling off as some of the cue ball's (neutron's) momentum is transferred or stolen from it. Thus, the proton flies off while the neutron is slowed down. Here particles like billiard balls are scattered off at angles which is not along the path of original cue ball which can be neutron.

There is a point where a beam of fast neutrons entering the moderator, "hydrogenous material", are slowed and scattered to such a high degree that it, the moderator, may appear as a neutron source of isotropically emitted "slow neutrons". These neutrons are said to be "thermalized".

Thermal neutrons are those neutrons that have a velocity equal to that of the atoms and molecules of their surrounding environment. They are in thermal equilibrium with all that is around them. All neutron counter detection schemes respond to thermal neutrons, so the need for moderators and neutron moderation exists. The detector is generally placed in the moderator such that any fast neutron beam that enters the moderator will thermalize to such a degree that the detector will be bathed in a sea of thermalized, easily counted, neutrons from all directions.

The moderator should, ideally, be a sphere for detection and measuring purposes, but this is not always practical. In general, a detector or material to be activated will need about 2-3 inches of the above moderating material surrounding it on all sides. Less thickness in a moderator will slow neutrons to "epi-thermal" velocities where a lot of activation materials have what are termed "capture resonances" to neutrons of energies just above thermal energies. These resonances can present monstrously huge cross sections for "neutron capture" within the material to be activated. This means that in some activation scenarios, a broad band of resonances and large capture cross sections might best be utilized for activation where the neutrons are at less than thermal energies.

If a moderator is made so large that the neutrons are slowed to velocities well below that of the surrounding molecules, they are termed "cold neutrons". For detection purposes, there is not much advantage for obtaining cold neutrons. However for activation studies cold neutrons, below thermal energy, increase the capture cross section and improve activation on a more or less linear scale based on the classic 1/V law. There is a limit, however, as the neutron flux is reduced in this effort due to increasing moderator volume and a trade-off point is reached for any given incident flux of fast neutrons outside the moderator.

Neutrons from fission have very high speeds and must be slowed greatly by water "moderation" to maintain the chain reaction. The uranium-235 is enriched to 2.5 - 3.5% to allow ordinary water to be the moderator (Figure 8.1).

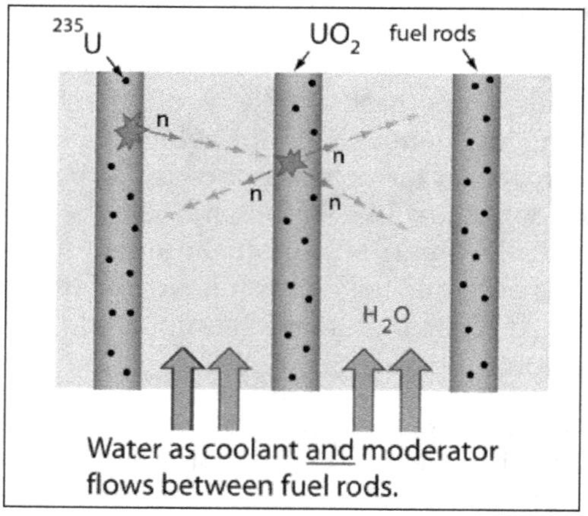

Figure 8.1: Water Moderator.

Loss of the water coolant kills the chain reaction since the fuel configuration is not "critical" without water moderation. Even with the moderator, the fuel is not "critical" without the inclusion of the "delayed" neutrons which may be emitted several minutes after the fission.

The water moderator is necessary in the uranium fission reactors. Loss of the water coolant kills the chain reaction since the fuel configuration is not "critical" without water moderation. Enough spontaneous fission events occur to to initiate a chain reaction if the proper moderation and fuel density is provided.

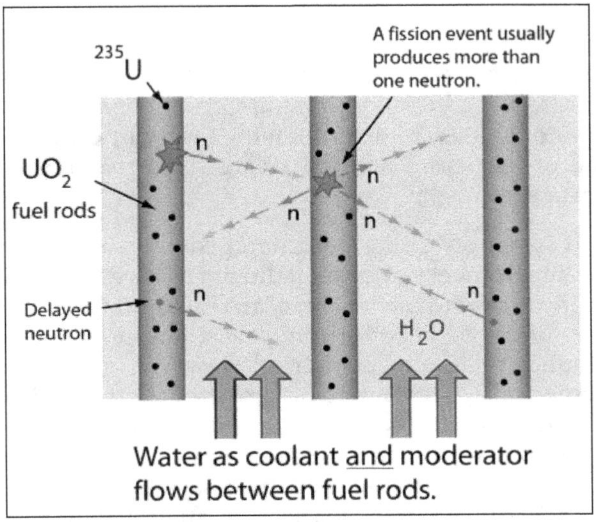

Even with the moderator, the fuel is not "critical" without the inclusion of the "delayed" neutrons which may be emitted several minutes after the fission. These neutrons come from the radioactive fragments from previous fissions.

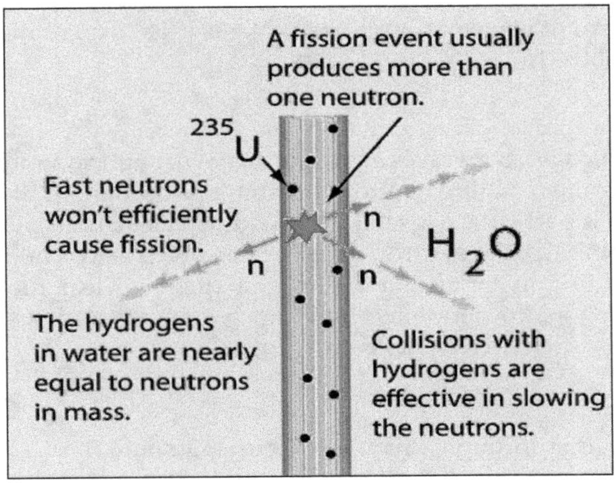

The neutrons emitted in nuclear fission reactions have high energies, typically in the range of 1 MeV. But the cross section for neutron capture leading to fission is greatest for neutrons of energy around 1 eV, a million times less. Neutrons with energies less than one electron volt are commonly referred to as "thermal neutrons" since they have energies similar to what particles have as a result of ordinary room-temperature thermal energy.

8.2 Neutron Transport in Fast Reactor Shields

8.2.1 Introduction

Neutron transport is the study of the motions and interactions of neutrons with materials. Nuclear scientists and engineers often need to know where neutrons are in an apparatus, what direction they are going, and how quickly they are moving. It is commonly used to determine the behavior of nuclear reactor cores and experimental or industrial neutron beams. Neutron transport is a type of radiative transport.

Neutron transport has roots in the Boltzmann equation, which was used in the 1800s to study the kinetic theory of gases. It did not receive large-scale development until the invention of chain-reacting nuclear reactors in the 1940s. As neutron distributions came under detailed scrutiny, elegant approximations and analytic solutions were found in simple geometries. However, as computational power has increased, numerical approaches to neutron transport have become prevalent. Today, with massively parallel computers, neutron transport is still under very active development in academia and research institutions throughout the world. It remains one of the most computationally challenging problems in the world since it depends on 3-dimensions of space, time, and the variables of energy span several decades (from fractions of meV to several meV). Modern solutions use either discrete-ordinates or Monte Carlo methods, or even a hybrid of both.

8.2.2 Types of Neutron Transport Calculations

Several basic types of neutron transport problems exist, depending on the type of problem being solved.

Fixed Source

A fixed source calculation involves imposing a known neutron source on a medium and determining the resulting neutron distribution throughout the problem. This type of problem is particularly useful for shielding calculations, where a designer would like to minimize the neutron dose outside of a shield while using the least amount of shielding material. For instance, a spent nuclear fuel cask requires shielding calculations to determine how much concrete and steel is needed to safely protect the truck driver who is shipping it.

Criticality

Fission is the process through which a nucleus splits into (typically two) smaller atoms. If fission is occurring, it is often of interest to know the asymptotic behavior of the system. A reactor is called "critical" if the chain reaction is self-sustaining and time-independent. If the system is not in equilibrium the asymptotic neutron distribution, or the fundamental mode, will grow or decay exponentially over time.

Criticality calculations are used to analyze steady-state multiplying media (multiplying media can undergo fission), such as a critical nuclear reactor. The loss terms (absorption, out-scattering, and leakage) and the source terms (in-scatter and fission) are proportional to the neutron flux, contrasting with fixed-source problems where the source is independent of the flux. In these calculations, the presumption of time invariance requires that neutron production exactly equals neutron loss.

Since this criticality can only be achieved by very fine manipulations of the geometry (typically via control rods in a reactor), it is unlikely that the modeled geometry will be truly critical. To allow some flexibility in the way models are set up, these problems are formulated as eigen value problems, where one parameter is artificially modified until criticality is reached. The most common formulations are the time-absorption and the multiplication eigen values, also known as the alpha and k eigen values. The alpha and k are the tunable quantities.

K-eigen value problems are the most common in nuclear reactor analysis. The number of neutrons produced per fission is multiplicatively modified by the dominant eigen value. The resulting value of this eigen value reflects the time dependence of the neutron density in a multiplying medium.

- keff < 1, subcritical: the neutron density is decreasing as time passes;
- keff = 1, critical: the neutron density remains unchanged; and
- keff > 1, supercritical: the neutron density is increasing with time.

In case of a nuclear reactor, neutron flux and power density are proportional, so during reactor start-up keff > 1, during reactor operation keff = 1 and keff < 1 at reactor shutdown.

8.3 Shielding for Variable Energy Cyclotron

A cyclotron is a type of particle accelerator invented in 1934 in which charged particles accelerate outwards from the centre along a spiral path. The particles are held to a spiral trajectory by a static magnetic field and accelerated by a rapidly varying electric field.

Cyclotrons were the most powerful particle accelerator technology until the 1950s when they were superseded by the synchrotron, and are still used to produce particle beams in physics and nuclear medicine. The largest single-magnet cyclotron was the 4.67 m synchrocyclotron built between 1940 and 1946 by Lawrence which accelerate protons to 730 MeV. The largest cyclotron is the 17.1 m (56 ft) multimagnet TRIUMF accelerator at University of British Columbia which can produce 500 MeV protons.

A spiral ridge cyclotron which can accelerate protons from 3.5 to 15 meV, deuterons from 7 to 14 meV, He^{3++} from 11 to 37 meV, and alpha particles from 15 to 27 MeV in which magnetic field is obtained from four spiral shims and five correction coils placed on each of the pole faces of the magnet. The spiral shims produce a maximum rms flutter of 0.07 which provides sufficient vertical focusing beyond a radial distance 6 in. from the center of the cyclotron. The two innermost correction coils provide magnetic focusing up to 6 in. and the three outer coils build up the isochronous field. Focusing effects of slits, placed on the dee and the dummy dee in the path of the particles for the first four revolutions, are demonstrated. The machine can deliver 1 ma of proton beam at the extraction radius, 18.5 in.

8.4 Computer Codes and Data Available from Radiation Shielding

Radiation can be a serious concern in nuclear power facilities, industrial or medical X-ray systems, radioisotope projects, particle accelerator work, and a number of other circumstances. Containing radiation and preventing it from causing physical harm to employees or their surroundings is an important part of operating equipment that emits potentially hazardous rays. Preserving both human safety and structural material that may be compromised from radiation exposure are vital concerns, as well as shielding sensitive materials, such as electronic devices and photographic film.

The process of regulating the effects and degree of penetration of radioactive rays varies according to the type of radiation involved. Indirectly ionizing radiation, which includes neutrons, gamma-rays, and X-rays, is categorized separately from directly ionizing radiation, which involves charged particles. Different materials are better suited for certain types of radiation than others, as determined by the interaction between specific particles and the elemental properties of the shielding material.

Almost any material can act as a shield from gamma or X-rays if used in sufficient amounts. Different types of ionizing radiation interact in different ways with shielding material. The effectiveness of shielding is dependent on the Stopping power of radiation particles, which varies with the type and energy of radiation and the shielding material used. Different shielding techniques are therefore used dependent on the application and the type and energy of the radiation. Shielding reduces the intensity of radiation depending on the thickness. This is an exponential relationship with gradually diminishing effect as equal slices of shielding material are added. A quantity known as the halving-thicknesses is used to calculate this.

A large set of computer codes and associated data libraries have been developed by nuclear research and industry over the past half century. A large number of them are in the public domain and can be obtained under agreed conditions from different Information Centres. The areas covered comprise: basic nuclear data and models, reactor spectra and cell calculations, static and dynamic reactor analysis, criticality, radiation shielding, dosimetry and material damage, fuel behaviour, safety and hazard analysis, heat conduction and fluid flow in reactor systems, spent fuel and waste management (handling, transportation, and storage), economics of fuel cycles, impact on the environment of nuclear activities *etc.* These codes and models have been developed mostly for critical systems used for research or power generation and other technological applications. Many of them have not been designed for accelerator driven systems (ADS), but with competent use, they can be used for studying such systems or can form the basis for adapting existing methods to the specific needs of ADS's.

Power reactors and research reactors operate at a multiplication factor of keff ≈ 1. The behaviour of a reactor in this status has been investigated very extensively, and reactors have been operated for several decades with a cumulative experience of many thousands of 'reactor years'. A large number of experiments were carried out in support of the design and operation of such reactors and were mostly tuned to the analysis of the critical state. Thus the nuclear data derived from experiments, the physics and the mathematical methods developed have focused to a very large extent to presently operating systems.

Accelerator driven systems (ADS) are operated at a sub-critical state and the most frequent value of keff considered in this case is 0.95. This choice is linked to criticality safety issues and regulations. It is well known that a sub-critical system, in order to sustain a continuous neutron flux requires an 'external' source as the neutrons released through fission are not sufficient to sustain a chain reaction due to the fact that the neutron loss is higher than its production. The 'external' source becomes thus an essential component of the system. The external sources can be quite different in nature: a neutron source such as AmBe, d-t , neutrons produced by proton (or hadron) accelerators through spallation, the photo-fission neutrons induced by intensive gamma fields from electrons beams hitting heavy targets, *etc.*

As per ADS study, preferred external neutron source is spallation source from proton accelerators, as of its high yield in neutrons. The sub-critical facility contains the spallation target made of material having high atomic mass that has a large number of excess neutrons available when the nucleus is split up in spallation residues. The spallation source is surrounded by a sub-critical configuration with the purpose of multiplying the neutrons produced by the source. The closer to criticality this surrounding system is, the higher the neutron multiplication. When increasing the keff from say 0.95 to 1.00 the neutron multiplication increases from 20 to 'infinity'. A critical reactor is thus a most powerful neutron multiplier.

The neutron transport equation as applied to fissionable systems also requires further analysis. The homogeneous radiation transport or Boltzmann equation, having 1/keff as the eigenvalue, is the one most widely known and used, but if used in connection with an external source requires a closer look to understand its meaning. This equation has as 'eigenvector' the fundamental mode observed in critical reactors. Keff is a convenient scale for criticality safety calculations as it is often the parameter used to determine safety margins. In a sub-critical system however the 'eigenvector' calculated represents a distribution in which the neutron production is artificially increased by the 1/keff factor, thus biasing the neutron distribution of the actual system. In a sub-critical system keff is kind of an artifice, for computing a reactivity scale. The homogeneous transport equation is still widely used, because of its convenience when using with existing computer codes and because of the familiarity many researches have with it. The actual neutron distribution in an external source driven system can be very different from the fundamental mode of the homogeneous balance equation. The shape of the external source is the more influential, the more the system is subcritical.

Neutron populations in strongly sub-critical systems remember where they come from and as we approach criticality due to the large multiplication and the increasing number of subsequent neutron generations the shape of the source is forgotten. Neutron populations in a critical system have no recall of the source (they have gone through a large number of generations), they populate the system with a distribution determined by the fundamental mode of the homogeneous critical equation.

The neutrons of an external source driven sub-critical system follow distributions made of super-positions of the fundamental and many higher modes, not of the homogeneous equation, but of the inhomogeneous equation containing the external source term. The Eigen values of the inhomogeneous equations are very different from the 1/keff of the homogeneous equation. The fundamental eigenvalue of the prompt neutrons is known as the α-Rossi, that is negative for systems below prompt critical. The fundamental Eigen values of the delayed neutrons are close but not identical to the delayed neutron precursor decay constant. The fundamental eigenvectors of prompt and delayed neutrons are not identical and differ from the one of the homogeneous equation with 1/keff.

As studied, Boltzmann equation including the delayed neutrons shows that there are large numbers of higher mode eigen values for both prompt and delayed neutron populations. The eigen values of the lower modes are all real, but for higher modes we find a continuum of eigenvalues in the complex domain.

The models we use in computer codes and the associated data libraries contain somehow the cumulative knowledge acquired during several decades of research. These do a good job if used by qualified staff in the parameter range for which they were validated and qualified. This was done by comparing the computed predictions with the results of 'integral' experiments, *i.e.* data describing some macroscopic quantities of the system. In essence our 'prediction' with computer code models is valid as long as it consists in 'interpolating' between sound knowledge we have accumulated. Calculations outside that range are 'extrapolations' into the unknown.

The results may be reasonable with some luck but we are moving in this case into 'forecasting' that may well be of the quality of the weather forecasting we see daily in the media. In essence, with ADS systems we move to some unknown ground, an area very interesting and challenging for research. But all new models developed for describing its 'neutronic'/'photonic' behaviour must be validated against experiments that cover the parameter space of ADS.

References

Blink. J. Choi. J. Farmer. J. 2007. Applications in the Nuclear Industry for Thermal Spray Amorphous Metal and Ceramic Coatings. Materials Science &

Technology 2007 Conference and Exhibition Detroit. MI. United States September 16.Volume 40.

Caldwell. Steven.. GWM Symposia.. 2013.. A Review of Tungsten Heavy Alloy Utilization in Isotope Transport Containers-13380 Material science Waste Menagemnet. United States.

Choi. J-S. Farmer. J.C. Lee. C. Fischer. L. M. July 42007. Neutron-Absorbing Coatings for Safe. Chemistry Materials and Life Sciences Directorate Lawrence Livermore National Laboratory UCRL-TR-232463 Volume 1.

Gilbert. MR. Sublet. JC., 2011. Neutron-induced transmutation effects in W and W-alloys in a fusion environment. Nuclear Fusion

Kibata. M.. Saito.Y.. Tsubota. M.. Doken. Y.. Sato.M.. 2013. Neutron shielding material. method of manufacturing the same. and cask for spent fuel. US8481986 B2 US 13/360.

Mawutorli Nyarku. Ramanathapura S. Venkata D., 2013 Experimental neutron attenuation measurements in possible fast reactor shield materials. Annals of Nuclear, 53, 135-139.

Mizia. RE. Lister. TE. Pinhero. PJ. Swank. WD.2007. Neutron absorbing coating for nuclear criticality control. US Patent 6919576.

Schmidt. E. Schmieden, 2013, Search for long-lived stopped R-hadrons decaying out of time with pp collisions using the ATLAS detector. Physical Review D 88, 112003.

Singh. V.P. and Badiger. N.M., 2014 Gamma-ray and neutron shielding properties of some alloy materials.AnnalsofNuclearEnergy, 301-310

Zhao. L. Robinson. L. Paul. RL.. 2007. Application of cold neutron prompt-gamma activation analysis in environmental studies of aquatic plants. Journal of Radioanalytical and Nuclear Chemistry. Volume 271. Number 3.

Index

www.ingramcontent.com/pod-product-compliance
Lightning Source LLC
LaVergne TN
LVHW011819200926
848484LV00005B/1243

* 9 7 8 1 7 8 7 1 5 0 1 5 7 *